Drawing
THE HEAD AND HANDS

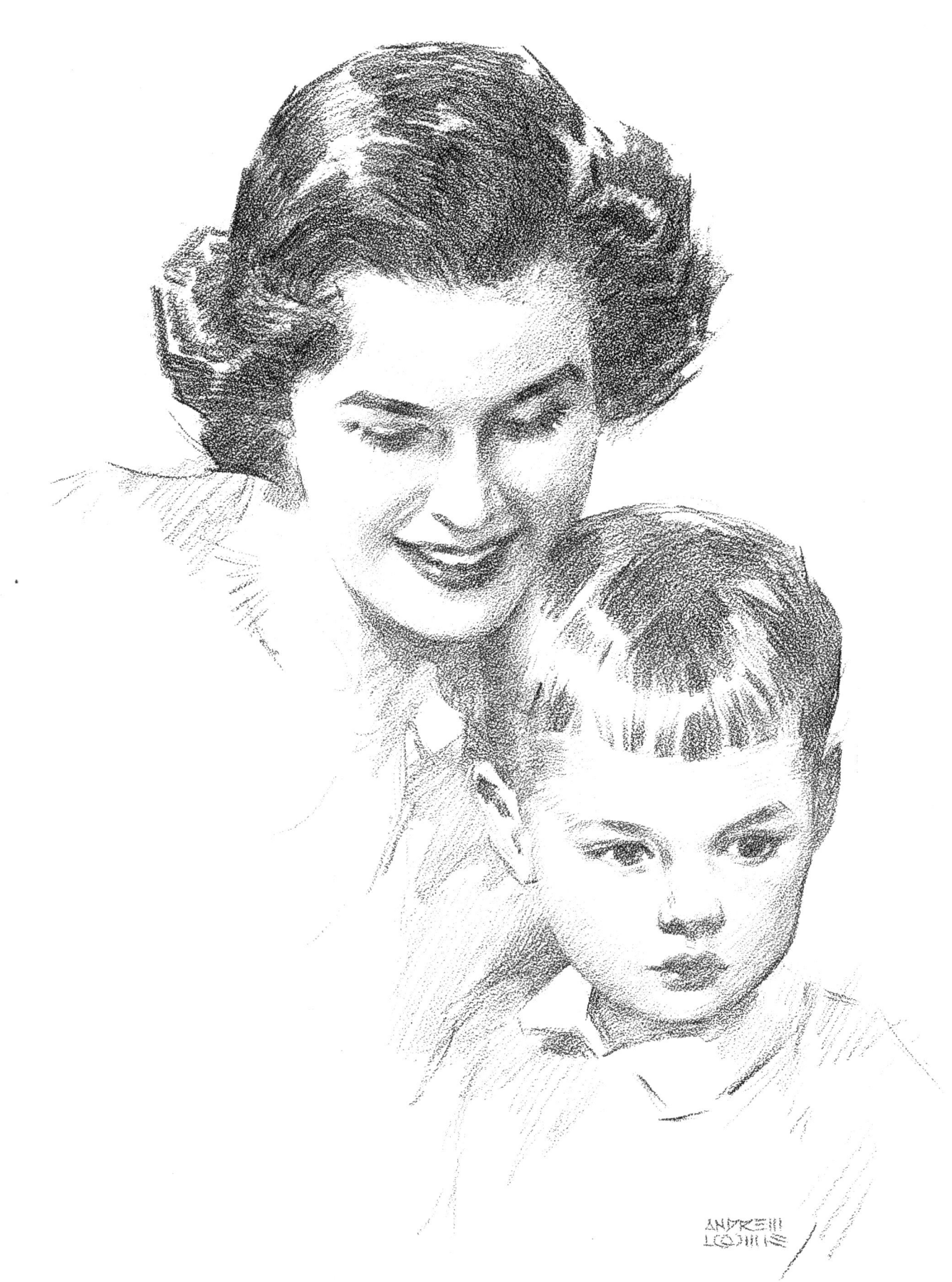

Drawing

THE HEAD AND HANDS

BY

ANDREW LOOMIS

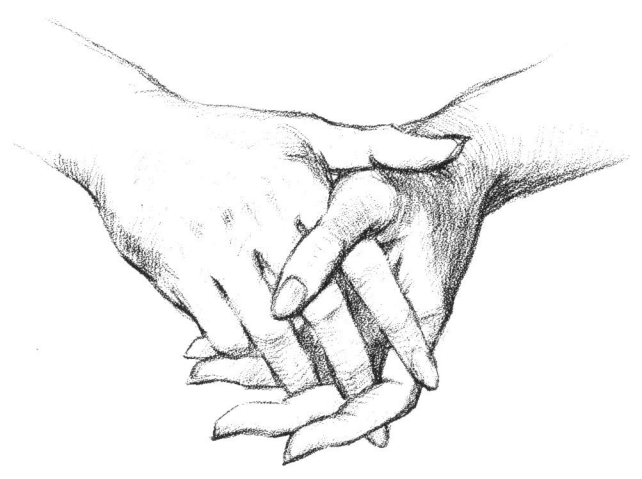

Drawing The Head And Hands
ISBN: 9780857680976

Published by
Titan Books
A division of Titan Publishing Group Ltd.
144 Southwark St.
London
SE1 0UP

This edition: October 2011
19 20

EU RP (for authorities only)
eucomply OÜ Pärnu mnt 139b-14 11317
Talinn, Estonia
hello@eucompliancepartner.com
+3375690241

To receive advance information, news, competitions, and exclusive offers online, please sign up for the Titan newsletter on our website: **www.titanbooks.com**

Did you enjoy this book? We love to hear from our readers. Please e-mail us at: **readerfeedback@titanemail.com** or write to Reader Feedback at the above address.

A CIP catalogue record for this title is available from the British Library.

Printed and bound in China.

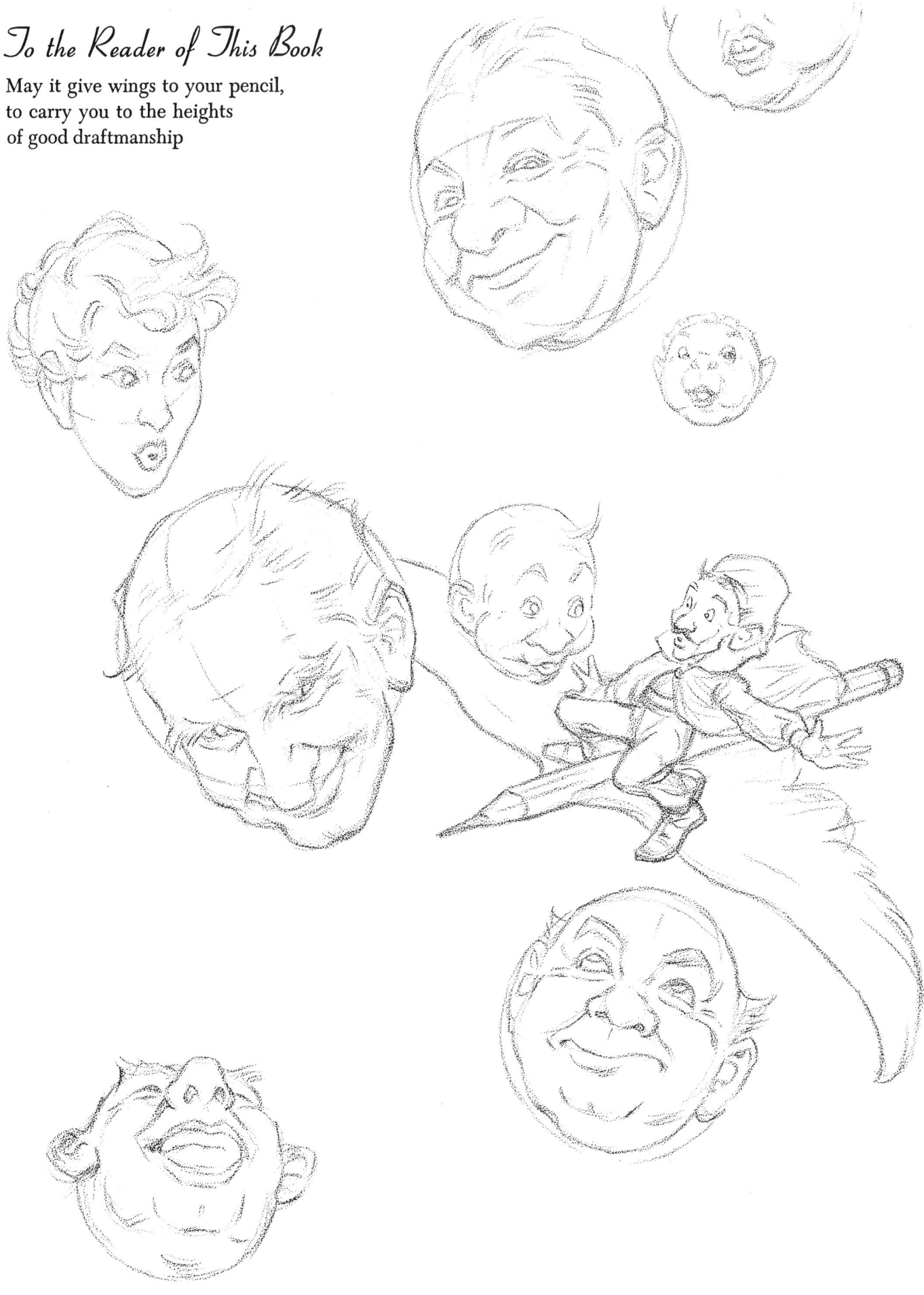

May I here express my appreciation for the valuable assistance in the preparation of this volume given me by my beloved wife, Ethel O. Loomis.

CONTENTS

(Illustration pages are indicated by italics)

CONTENTS

Drawing

THE HEAD AND HANDS

A Short Chat with the Reader

How FORTUNATE it is for the human race that every man, woman, and child is tagged with an individual and identifiable face! If all faces were identical, like the labels on a brand of tomatoes, we would be living in a very mixed-up world. When we think of it, life is mainly a continuous flow of experiences and contacts with people, different people. Suppose for a moment that Jones, the egg man, was the exact counterpart of Smith, the banker; that the face across the table might be that of Mrs. Murphy, Mrs. Goldblatt, or Mrs. Trotsky—you couldn't be sure which. Suppose all the faces in the magazines and newspapers and on television were reduced to one male and one female type, what a dull thing life would be! Even if your face has not been your fortune, even if it is far from beautiful, still, nature really gave us all a pretty good break, for at least we are individuals and can each be thankful for having a face, good or bad, that is undeniably our own.

This individuality of faces can be an intensely interesting study for anyone, and especially for anyone with the slightest talent for drawing. Once we begin to comprehend some of the reasons for the differences, our study becomes all-absorbing. Through our faces, nature not only identifies us but tells the world a good deal more about each of us.

Our thoughts, our emotions and attitudes, even the kind of lives we live, register in our faces. The mobility of the flesh—that is, the power of expression—adds more than mere identity. Let us give more than casual attention to the endless procession of faces moving in and out of our consciousness. Setting aside the psychological and emotional phases of expression, we can express in simple language the basic technical reasons for the smile, the frown, and all the variations that we call facial expression. We say that a person can look guilty, ashamed, frightened, content, angry, smug, confident, frustrated, and a host of other ways too numerous to tabulate. A few embedded muscles attached to the bones of the skull provide the mechanics for every expression, and these muscles and bones are not complicated or difficult to learn! What a wealth of interest lies within!

Let me say at the beginning that to draw a head effectively is not a matter of "soul searching" or mind reading. It is primarily a matter of interpreting form correctly in its proportion, perspective, and lighting. All other qualities enter the drawing as a result of the way that form is interpreted. If the artist gets that right, the soul or character is revealed. As artists, we only see, analyze, and set down. A pair of eyes drawn constructively and in correct values will appear to be alive because of craftsmanship, not because of the artist's ability to read the sitter's soul.

The element that contributes most to the great variation of identities is the difference in the shapes of the skull itself. There are round heads, square heads, heads with wide and flaring jaws, elongated heads, narrow heads, heads with receding jaws. There are heads with high domes and foreheads, and those with low. Some faces are concave, and others convex. Noses and chins are prominent or receding. Eyes are large or small, set wide apart or close together. Ears are all kinds of shapes and sizes. There are lean faces and fat faces, big-boned and small-boned ones. There are long lips, wide lips, thin lips, full lips, protruding lips, and equal variety in the sizes and shapes of noses. You can see that, by cross multiplication of these varying factors, millions of different faces will be produced. Of course, by the law of averages certain combinations of factors are bound to reappear. For that reason people who are not related sometimes closely resemble each other. Every artist has

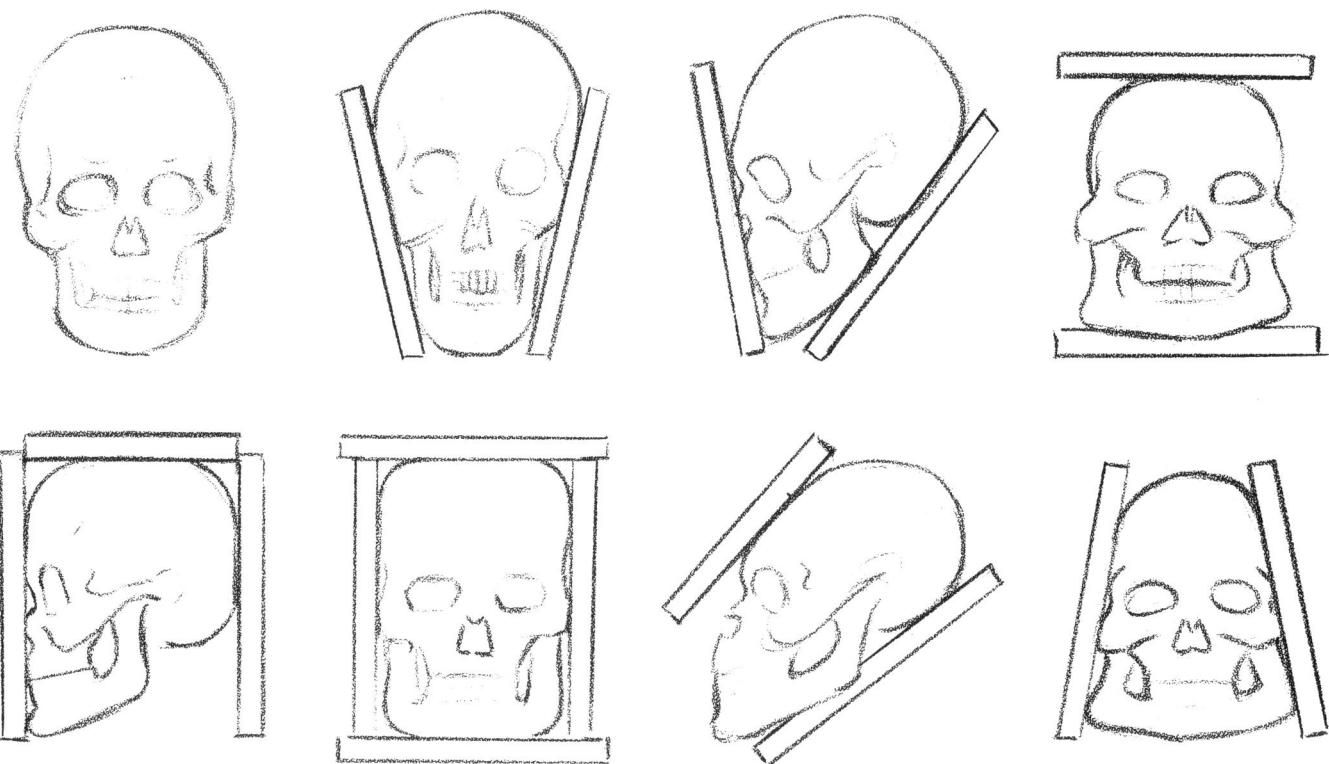

had the experience of being told by someone that a head he has painted or drawn looks like that person or like an acquaintance or relative of the speaker.

For the artist's purpose, the simplest plan is first to think of the skull as being pliable and having taken a certain shape as a result of pressures—as if one squeezed a rubber ball into various shapes without changing its actual volume. Although skulls have a great variety of shapes, actual measurements tally very closely, which means that the volume is about the same and only the shape is different. Suppose we model a skull in soft clay, then, between boards, press it into various shapes. Thus out of the same volume we can make a narrow head, a wide head, flaring jaws, and all the other types. How heads got to be this way is not our problem, which is only to analyze and thus determine the type of skull in the particular head we wish to draw. Later, when you become more familiar with the construction of the skull, you will be able to show these variations so successfully that you will be able to draw practically any type you choose and make it convincing.

At the same time you can set down understandingly any type before you. By the time you understand how the flesh is distributed over the bones of the face, you will be able to vary the expression of the same head. The thing to remember is that the skull is fixed in position, and, with the exception of the jaw, immovable, and that the flesh is mobile and ever-changing, and also affected by health, emotion, and age. After the skull is fully matured, it remains the same through life and is a structural foundation for the varying appearance of the flesh. Therefore the skull is always the basis of approach, and all other identifying features are built into or upon it.

From the skull we get the spacing of the features, which is more important to the artist than the features themselves. The features must take their proper places in our construction. If they do, we have little trouble in drawing them. Trying to draw the features without having located them properly is an almost hopeless task. Eyes do strange things; mouths leer instead of smile; faces take on weird and unholy expressions. In trying to correct a face that appears to be out of drawing, the chances are that we will do just the wrong thing. Instead of moving an eye into its socket, we trim down a cheek; if a jaw line is out, we add more forehead. We should know, in first laying in the outline, that the whole head is in construction. This I am sure you can learn from the pages that follow.

The big difference between the completely amateur attempt and the well-grounded approach is that the beginner starts by setting eyes, ears, noses, and mouths into blank white space, surrounded by some sort of an outline for the face. This is drawing in the two dimensions of height and width only. We must somehow get into the third dimension of thickness, which means that we must draw the whole head as it exists in space and build the face upon it. By doing so we are able not only to place the features, but also to establish the planes of light and shadow, and, further, to identify the humps, bumps, and creases as being caused by the underlying structure of muscle, bone, and fat.

To help the beginner to start out with this third dimension, many approaches are suggested by various teachers. Some use an egg shape; others a cube or block. Some even start with one feature and start building the form out around it until the whole head is encompassed. However, all these involve many chances for error. Only the front view of the head looks like an egg, and even that gives no line of the jawbone. In profile the head is not like an egg. As for the cube, there is no accurate way of setting the head into it. The head is totally unlike a cube from any angle. The only value the cube has in drawing heads is to help set the construction lines into perspective, as you will learn later.

It seems more logical to start with a shape that is basically like the skull, one that is simple to draw and is accurate for purposes of construction. This can be done by drawing a ball resembling the cranium, which is round but flattened somewhat at the sides, and attaching the jawbone and features to it. Some years ago I hit upon this plan and made it the basis of my first book, *Fun with a Pencil*. I am happy to say that the plan was received with great enthusiasm and is now widely used in schools and by professional artists. Any direct and efficient approach must presuppose the skull and its parts and its points of division. It is just as reasonable to start drawing a wheel with a square as it is to start drawing a head with a cube. By cutting off corners and further trimming the square you could eventually come out with a fairly good wheel. You could also chip away the cube until you had a head. But at best it's a long way around. Why not start with the circle or ball? If you can't draw a ball, use a coin or a compass. The sculptor starts with a form of the general shape of the face attached to the ball of the cranium. He could not do otherwise.

I present this simple plan in this volume since it is the only approach that is at the same time creative and accurate. Any other accurate approach requires mechanical means, such as the projector, tracing, the pantograph, or using a squared-off enlargement. The big question is really whether you wish to develop the ability to draw a head, or whether you are content to use mechanical means of projecting it. My feeling is that, if the latter were the case, you would not have been interested in this book. When your bread and butter depends upon creating an absolute likeness, and you do not wish to gamble, make the best head you can by any means possible. However, if your work is to give you joy and the thrill of accomplishment, I urge you to aim at the advancement of your own ability.

The drawings on pages 14 and 15 show the possibilities of developing all kinds of types out of the variations of skulls. After you have learned to set up the ball and plane, you can do almost anything you please with it, fitting all parts into the construction by the divisions you make across the middle line of the face. You have at your disposal jaws, ears, mouths, noses, and eyes, all of which may be large or small. The

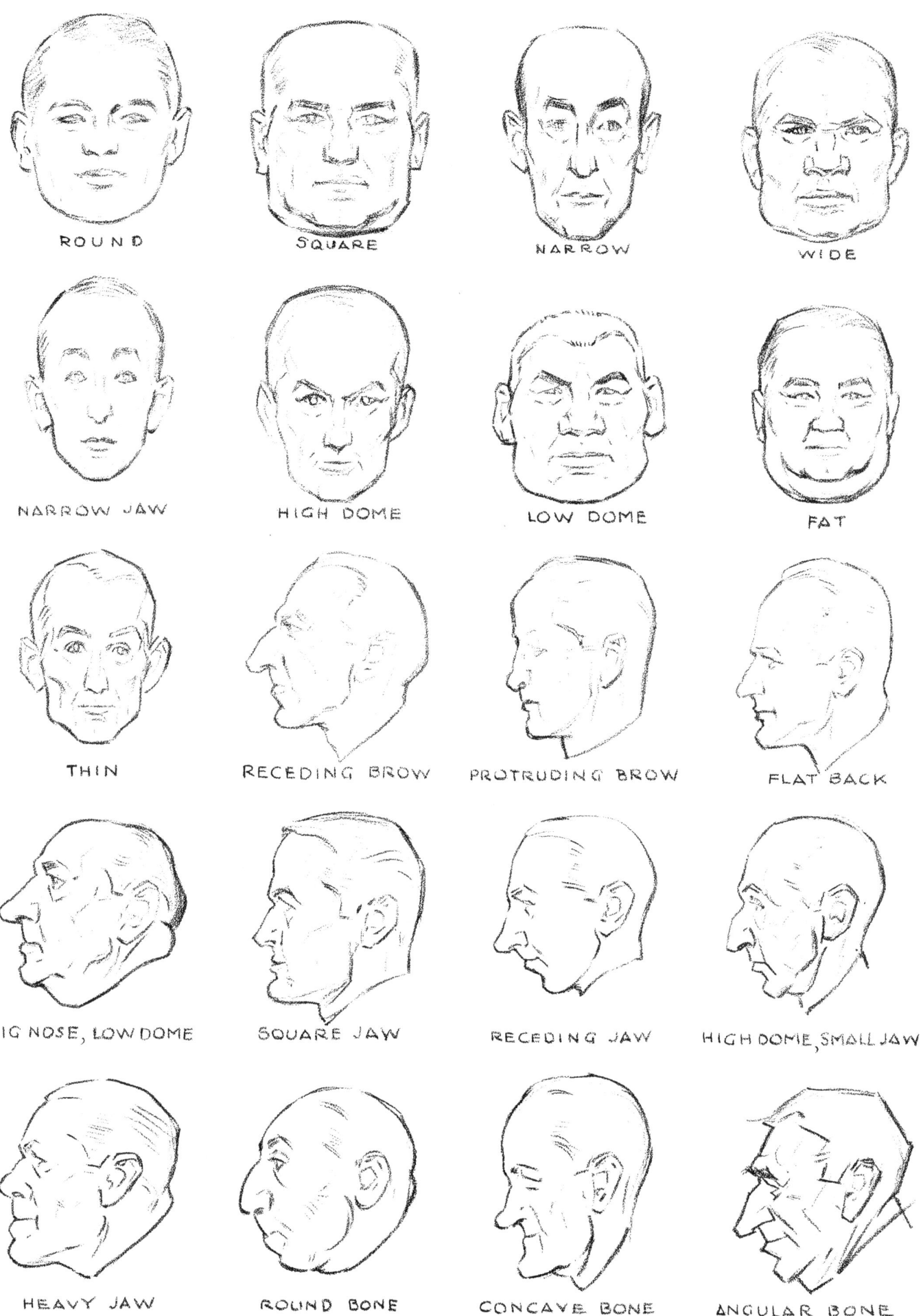

ROUND   SQUARE   NARROW   WIDE

NARROW JAW   HIGH DOME   LOW DOME   FAT

THIN   RECEDING BROW   PROTRUDING BROW   FLAT BACK

BIG NOSE, LOW DOME   SQUARE JAW   RECEDING JAW   HIGH DOME, SMALL JAW

HEAVY JAW   ROUND BONE   CONCAVE BONE   ANGULAR BONE

14

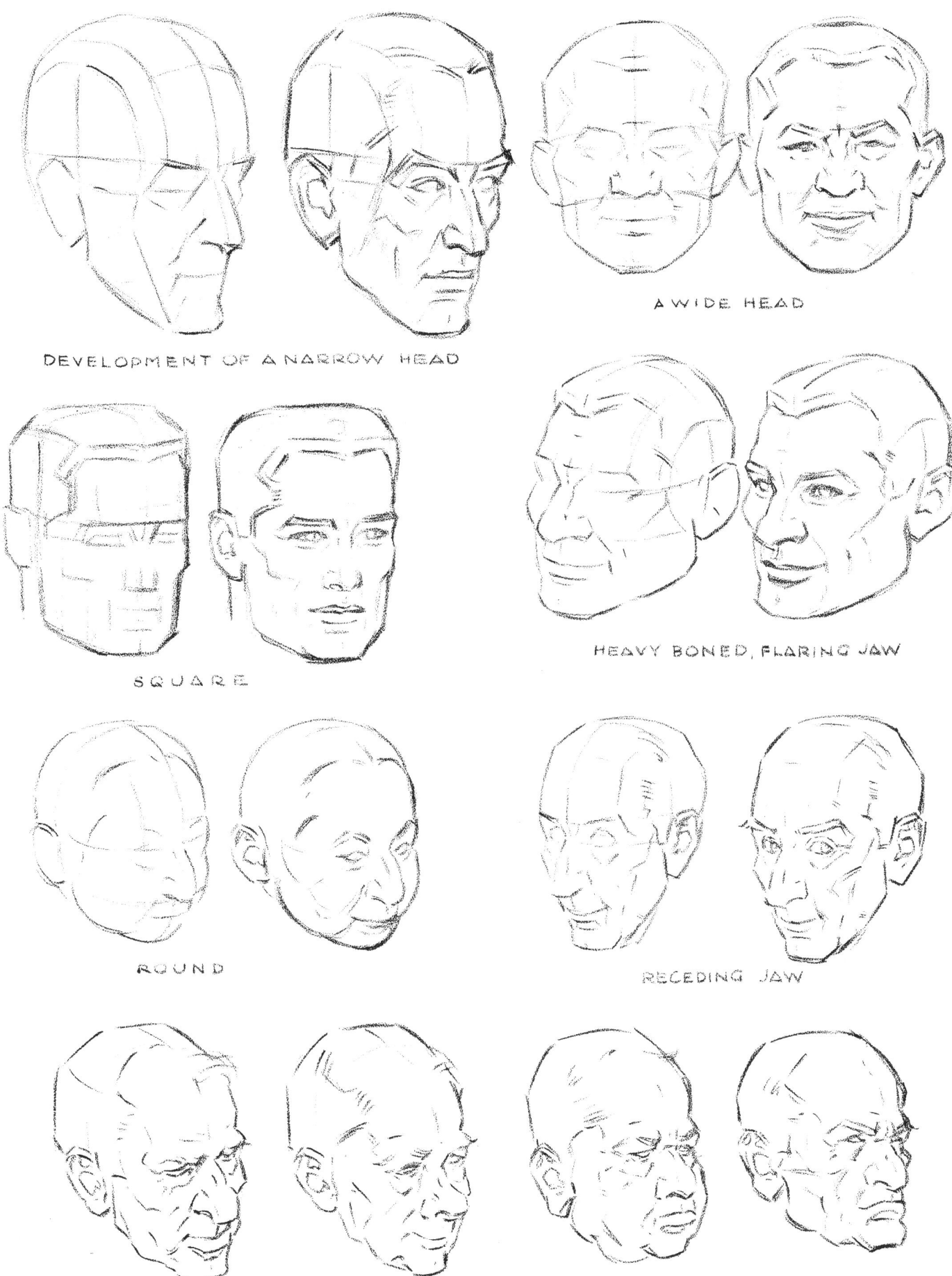

DEVELOPMENT OF A NARROW HEAD

A WIDE HEAD

SQUARE

HEAVY BONED, FLARING JAW

ROUND

RECEDING JAW

DIFFERENT FEATURES ATTACHED TO THE SAME CRANIUM

15

cheekbones may be set high or low, the upper lip may be long or short, the cheeks full or sagging. By different combinations of these, you can produce an almost endless variety of characters. It will be great fun for you to experiment.

Although the construction of any head involves more or less the same problems, this book is divided into sections on drawing men, women, and children of various ages. As we shall see, though the technical differences are slight, there is considerable difference in approach and feel-ing. The technical problems are explained in Part One, and the knowledge acquired from that is applied in the later sections on heads.

To be able to draw hands convincingly is also very important to the artist, and in this field too there is little material available. So Part Five has been included to help you understand the principles of construction on which realistic rendering of hands must be based.

Now let's get to work in earnest.

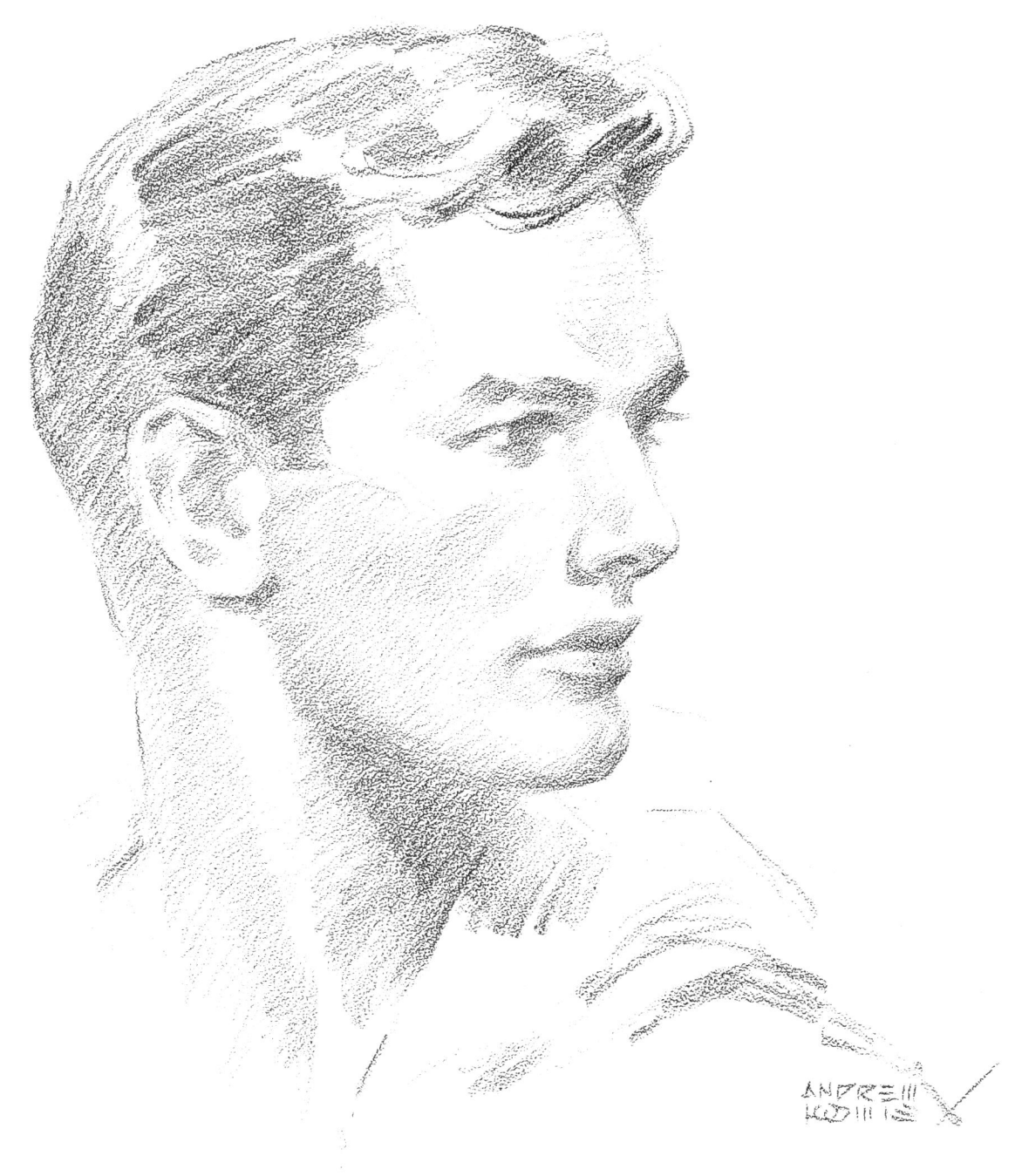

Part One: Men's Heads

Part One: Men's Heads

LET US BEGIN by establishing our common objective. You may be interested in drawing as a hobby. You may be an art student attending drawing classes. You may be a young professional, out of school, striving to better your work so that it will bring in more income. Perhaps you studied art many years ago and now have the time and incentive to take it up again. Perhaps you are well established in the field of commercial art, where competition is formidable, and are looking for something that will help you hold your place and, if possible, keep you moving forward. Whichever category you are in, this book will be helpful to you, because it provides practical knowledge of the techniques of drawing heads, both for the complete beginner and to help the more advanced artist in those most frustrating moments when the head he is drawing seems to refuse to do his work justice.

There must be a genuine basic motive behind any genuine effort. Ask yourself quite honestly, "Why do I really want to draw heads and draw them well?" Is it for the satisfaction of personal accomplishment? Does it mean enough to you to give up time from other things in order to learn? Do you hope someday to sell your work and make it your means of livelihood? Would you like to draw portraits, girls' heads for calendars, illustrations for magazine stories, the people in advertisements? Do you want to improve your drawing of heads to help sell your work? Is drawing a form of relaxation to you, helping to relieve tension and clear your mind of worries and other problems? Search quietly and thoroughly for this basic motive, because if it is powerful enough, it will give your efforts the strength to withstand discouragement, disappointment, disillusionment, or even seeming failure.

May I add one suggestion? Whatever your motive, try not to be impatient. Impatience has probably been a bigger stumbling block in the way of real ability than anything else. Doing anything well, I'm sure, means hurdling obstacles of one kind or another most of the way to the goal. Skill is the ability to overcome obstacles, the first of which is usually lack of knowledge about the thing we wish to do. It is the same in anything we attempt. Skill is a result of trying again and again, applying our ability and proving our knowledge as we gain it. Let us get used to throwing away the unsuccessful effort and doing the job over. Let us consider obstacles as something to be expected in any endeavor; then they won't seem quite so insurmountable or so defeating.

Our procedure will be a little different from that of the usual textbook. In general, textbooks seem to confine the material solely to problem and solution, or to technical analysis. That, in my own belief, is one of the reasons why textbooks are so difficult to read and digest. Every concentrated creative effort involves a personality, since skill is a personal matter. Since we are dealing not with organic material like nuts and bolts, but with human qualities like hope and ambition, faith or discouragement, we must throw out the textbook formulas and consider personal achievement as the basic element of our planning. An instructor would not be very helpful if he gave his students only the words of a textbook, all cold hard fact, without feeling, without praise or personal encouragement. I cannot participate in all your personal problems, but I can certainly remember my own, and assume that yours will not be greatly different. Therefore this book anticipates the solution of these problems even before you meet them. I believe that is the only way to handle this type of subject effectively.

There is an element of joy in doing what you

have proved to yourself to be right. It is my job here to give you the working materials with which to make your own effort successful rather than to show that anyone can succeed. Success comes only with personal effort, aided by whatever knowledge the individual can apply along with the effort. If this were not true, we would be able to do anything in the world simply by reading books. We all know this is not true. There are books on almost any subject. Their value depends upon the amount of knowledge they contribute and on how well it is absorbed and put into practice.

To draw heads well, the artist must detach his mind from the sitter's emotional qualities and develop an objective viewpoint. Otherwise he could go on drawing the same head forever, almost each moment noting a subtle change of expression, or a different mood in the subject. One face can vary in a thousand ways, and a drawing must show the effect of a single instant. Let him think of the head as only so much form in space, like a piece of still life rather than as an ever-changing personality.

To the beginner there is a certain advantage in drawing from a cast, or from a photograph, for at least the subject is not moving and he can regard it objectively. It is logical that our book begin purely from an objective approach with a form most like the average head, with average features and average spacings. Individual characteristics are much too complicated until we are able to tie them into a basic structure, one that is reasonably sound and accurate. Let us fix in our minds that the skull itself is the structure and all the rest merely trimmings.

Anatomy and construction can appear dull, but not to the builder. It might be dull to learn how to use a saw and hammer, but not when you are making a building of your own. It may be hard to think of the head as a mechanism. But if you were inventing a mechanism, it would never lack interest. Just realize that the head must be a good mechanism in order to be a fine head, and you will draw it with as much interest

as you would have in putting a part into a motor which you wanted to give a good performance.

It is evident, then, that we need to start with a basic shape that is as nearly like the skull as we can get it. Looking at the cranium, we see it most nearly like a ball, flattened at the sides and somewhat fuller in the back than the front. The bones of the face, including the eye-sockets, the nose, the upper and lower jaw, are all fastened to the front of this ball. Our first concern is to be able to construct the ball and the facial plane so that they operate as one unit which may be tipped or turned in any manner. It is of utmost importance that we construct the head in its complete and solid form, rather than just the visible portion of it. Naturally we cannot see more than half the head at any time. From the standpoint of construction, the half we cannot see is just as important as the visible half.

If you look at Plate 1, you will note that I have treated the ball as if the under half were transparent so that the construction of the whole ball is made evident. In this way the drawing on the visible side of the head can be made to appear to go all the way round, so that the area we cannot see can be imagined as a duplicate of what we do see. An old instructor of mine once said, "Be able to draw the unseen ear," which, at the time, puzzled me no end. I later realized what he meant. A head is not drawn until you can feel the unseen side.

It must be obvious from the preceding that it is impossible to draw the head correctly by starting with an eye or nose, oblivious of the skull and the placement of features within it. One might as easily try to draw a car by starting with the steering wheel. In all drawing no part can be as important as the whole, and the whole is always a fitting together of proportionate parts. We can always subdivide the whole into its parts, instead of guessing at the parts, hoping they will go together in the proper proportions. For example, it is easier to know that the forehead is one-third of the face, and what its position is on the skull, than to build the skull from

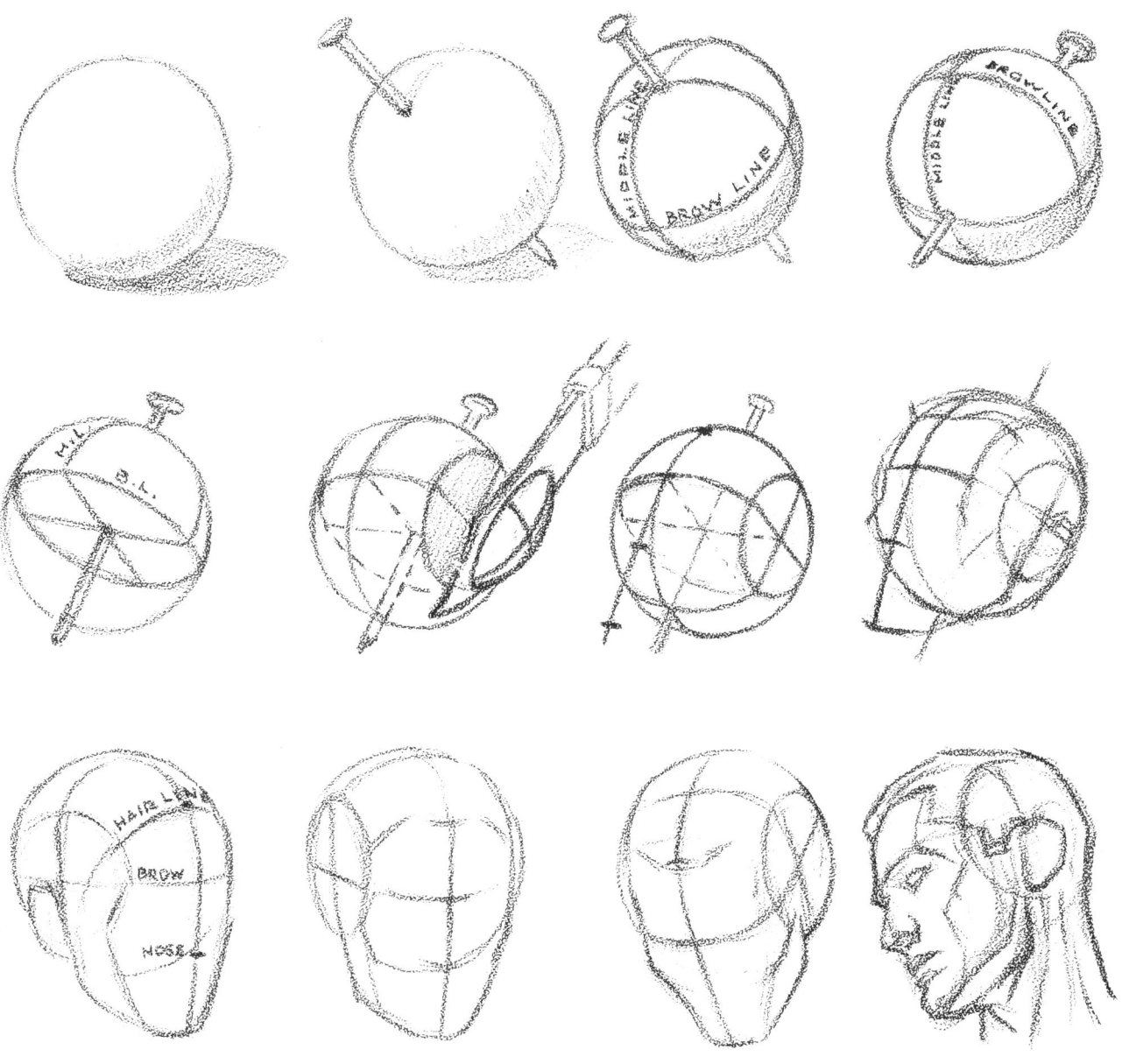

PLATE 1. The basic shape is a flattened ball

The cranium is more like a ball than anything else. To represent the ball as a solid sphere, we must establish an axis, like the nail through the ball at the top. Through the centers established by the axis, we can divide the ball into quarters and again at the equator. Now if we were to slice off a fairly thin slice on each side, we will have produced a basic shape that very closely matches the cranium. The "equator" becomes the brow line. One of the lines through the axis becomes the middle line of the face. About halfway up from the brow line to the axis, we establish the hairline, or the top of the face. We drop the middle line straight down off the ball. On this we mark off two points about equal to the space of the forehead, or from brow line to hairline. This gives us the length of the nose, and below that the bottom of the chin. We can now draw the plane of the face by drawing in the jaw line, which connects about halfway around the ball on each side. The ears attach along the halfway line (up and down) at a distance about equal to the space of from the brows to the bottom of the nose. The ball can be tipped in any direction.

21

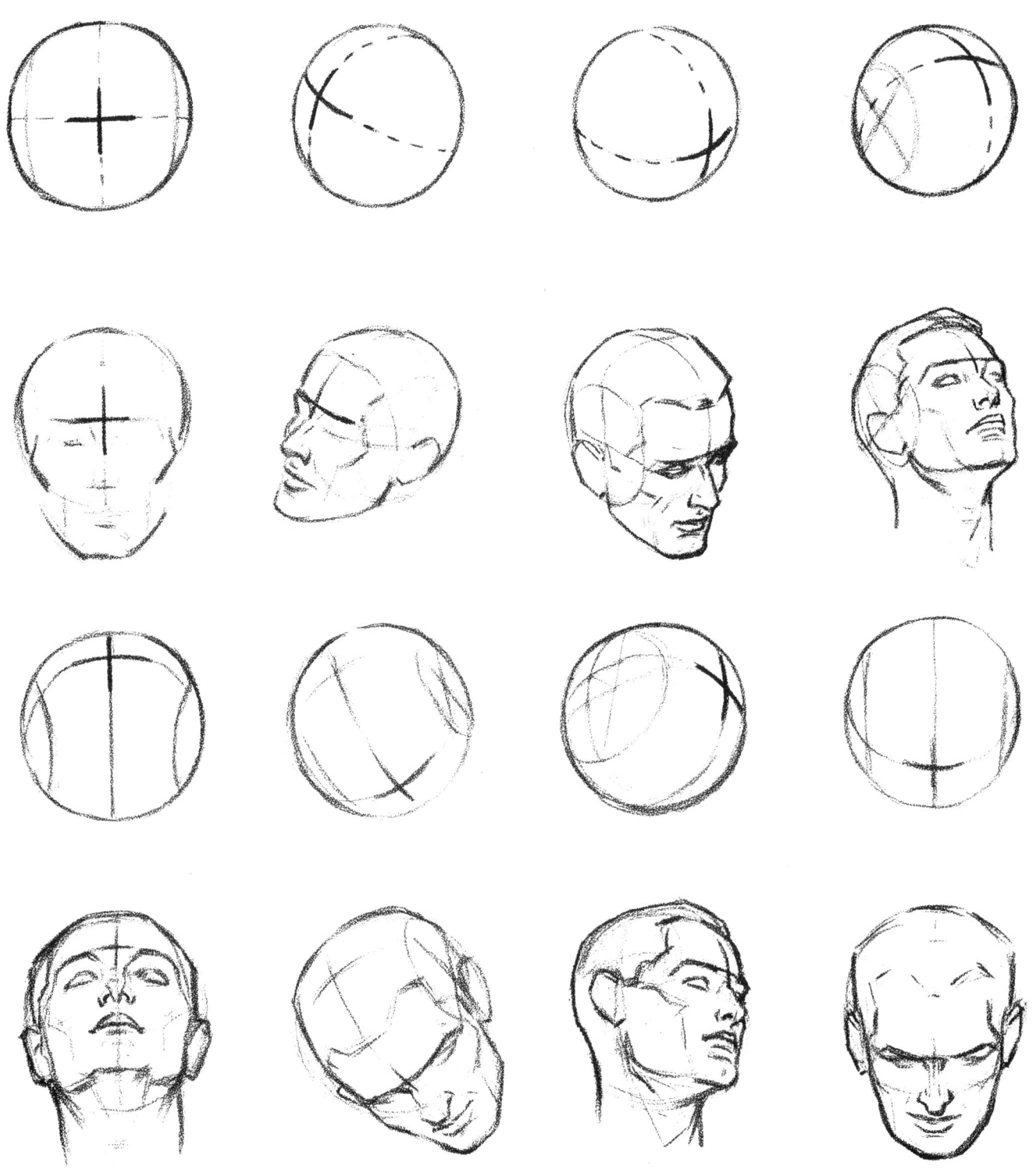

PLATE 2. The all-important cross on the ball

The "cross," or the point where the brow line crosses the middle line of the face, is the key point of the construction of the whole head. It determines the position of the facial plane on the ball, or the angle from which we see the face. It is easily spotted on the model or copy. By continuing the line up and down, we establish the middle line of the whole head. We draw the two sides of the face and head from this line. By continuing the brow line around the head we can locate the ears.

22

the forehead. Perhaps we have always thought of the head so much in terms of belonging to a definite individual that we have never considered it in a mechanical sense. It perhaps never occurs to us that a smile is a mechanical principle in action, as well as evidence of a beaming personality. Actually the mechanics involved in a smile are the same as those used in a drawstring on a curtain. The string is attached to something fixed at one end, and to the material at the other. Pulling the string buckles the material. The cheek plumps out in the same way. The working of the jaw is like a hinge or a derrick, but the hinge is of the ball-and-socket type. The eyes roll in their sockets like a ball bearing held in place. The eyelids and the lips are like slits in a rubber ball, which naturally close except when they are pulled apart. There is a mechanical principle beneath every expression put into action by the brain. Underlying the flesh of the face are muscles which are capable of expansion and contraction, just like all the other muscles of the body. We discuss this interesting material in more detail later.

We start drawing the head by establishing points on the ball and on the facial plane. Both the ball and the facial plane must be subdivided in order to establish those points. No matter how much you draw, how skilled you get to be, how well trained your eye becomes, you will always have to begin by building the head correctly, just as a carpenter, no matter how long he has worked, always measures a board before he cuts it. Construction of the face and head depends upon establishing the points of measurement. Any other way is bound to be guesswork, which is a gamble any way you take it. For the one time you guess right, there are many inevitable mistakes.

The most important point in the head from which to build the construction of the face is the point immediately above the bridge of the nose, between the brows. This point remains always fixed and is indicated by the vertical line of the nose and the crossline of the brows. On

the ball this is the junction of the "equator" and "the prime meridian," the two lines that cut the ball in half vertically and horizontally. All measurements spring from this point. About halfway up from this point to the top center of the head we get the hairline, and have therefore spaced off the forehead. Dropping down an equal distance below the crosspoint, we get the length of the nose, since the distance from the tip of the nose to the brows is, on an average, equal to the height of the forehead. Measuring the same distance down, we get the bottom of the chin, for the distance from the bottom of the chin to the base of the nose equals the space from there to the brows, and from that point to the hairline. So it's one, two, three spaces, all equal, down the middle line of the face. See Plates 3 and 4. I suggest you take paper and pencil and start drawing these heads, tipping them in every possible direction. This can well be your first real period of study. What you do now will affect everything you do from here on. Plate 4 will give you an idea of how to place the features properly. The placement is more important than the drawing of the features themselves. At this stage it is not too important that the details of the features be correct. Get them to fall within the construction lines, so that the two sides of the face seem to match, whatever the viewpoint.

The next time you work with this book, turn to Plate 5, which is a simplified statement of the bone structure. No one detail of the bone structure is of great importance, but its total shape is of paramount importance. Within the shape we must locate the eye-sockets, spacing them carefully on either side of the middle line. We locate the two cheekbones opposite each other, and the bridge of the nose, which must lie on the middle line at the top and extend out from the middle line at the bottom. We locate the corner of the jaw and bring the jaw line down to the chin. Every head must be constructed so that all the features balance on the middle line. Plate 6 gives you more of the actual appearance and placement of the bones. Note how in these drawings you are aware of the construction all around the head. I personally try to get the feeling that these are not outlines, but the edges of solid forms that I could slide my hand around. Do you feel as if you could pick up these heads with your two hands and that you would find them just as solid in back as in front? That is what we are working for just now.

Plate 7 shows the action of the head on its pivot point at the top of the spine and at the base of the skull. We must remember that this pivot is well inside the roundness of the neck and deep under the skull. It does not have a hinge action but a rotating action from a point a little back of the center line of the neck. So when the head is tipped backward the neck is squeezed and bulges somewhat, forming a crease at the base of the skull. When the head is tipped forward, the larynx or Adam's apple is dropped down and hides itself within the neck. In the lateral movements there is a strong play of the long muscles which attach to the skull behind the ears and down in front to the breastbone between the collarbones. At the back are the two strong muscles which attach to the base of the skull to pull the head backward. To get a head to sit properly on the neck requires some knowledge of anatomy, which is discussed later.

Some artists like to think of the head as being built of pieces which will fit together and fall into place to give the understructure of the head. See Plate 8. This is especially helpful in suggesting the third dimension, that of thickness, in your drawing. Much too often the face is drawn as something flat. We must consider the roundness of the muzzle—the two jaws as they come together. Because it is lost in the fleshiness of the face, we may forget the sharp curve of the teeth behind the lips. This is even more pronounced in animals, to which a sharp deep bite may make the difference between life and death. Think of the front teeth as choppers and the back teeth as grinders. The fangs, or what we call eyeteeth in human beings, are what an animal uses to hang on with, or to slash and tear. To impress upon yourself what the roundness of this area really is like, take a bite out of a piece of bread and study it. You will probably never draw lips flatly again. We must also remember that the eyes are round, though most of the time we see them drawn flatly, like a slit in a piece of paper. The eyes, nose, mouth, and chin all have this three-dimensional quality, which cannot be sacrificed without losing the solidity of the whole head.

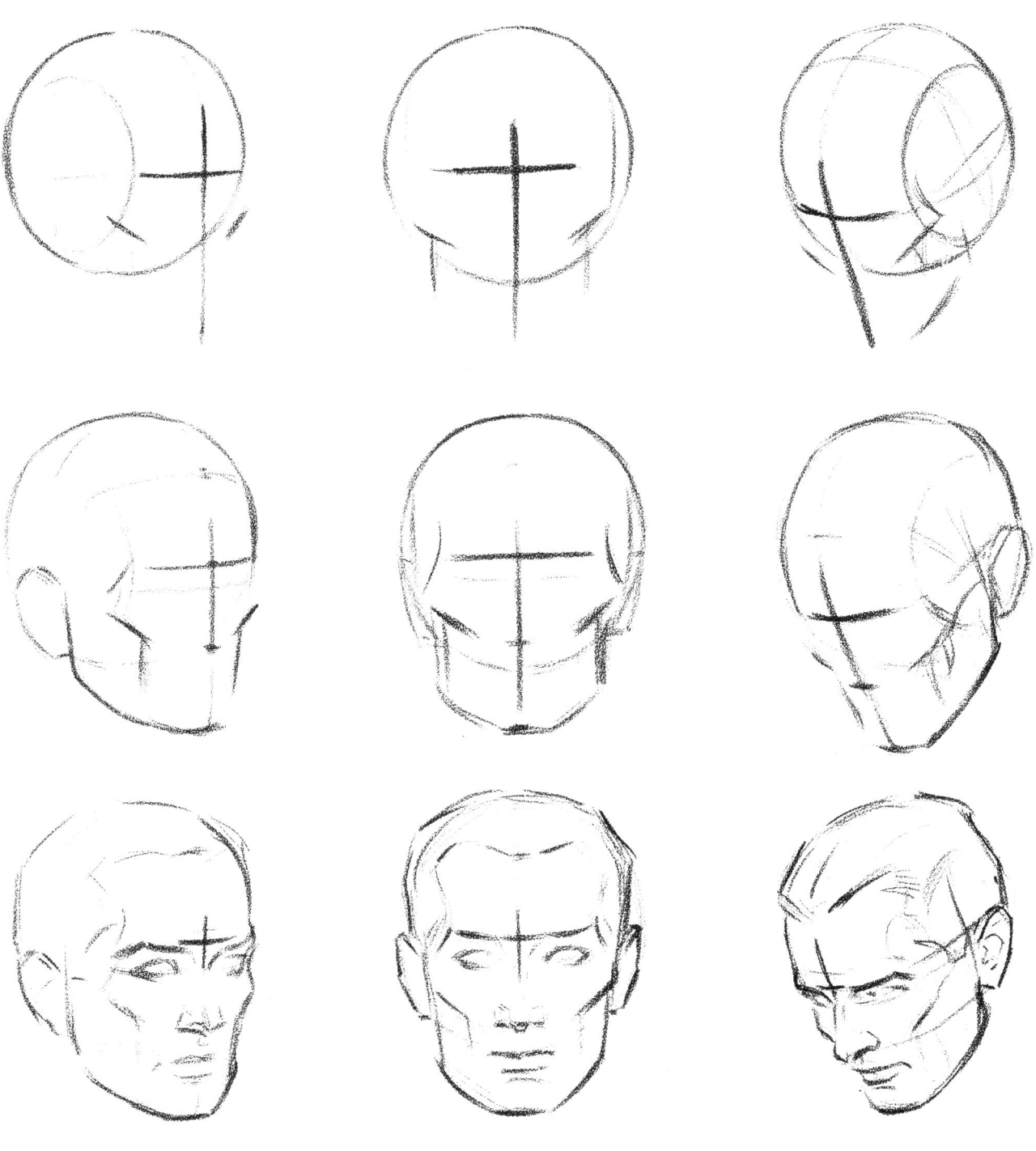

PLATE 3. The cross and the middle line determine the pose

Get out your pencil and pad.

It is most important to begin at once to practice setting up the ball and facial plane. Do not worry too much now about the features. This is simply construction, which you will probably use for the rest of your life. Establish the cross. Try to think of the construction all around the head, so that the jaws attach halfway around on each side. Remember that the eyes and cheekbones are below the brow line. The ears are about parallel with the lines of the brows and that of the nose. The cross almost suggests the face below. With this approach we can start drawing the whole head in any pose.

25

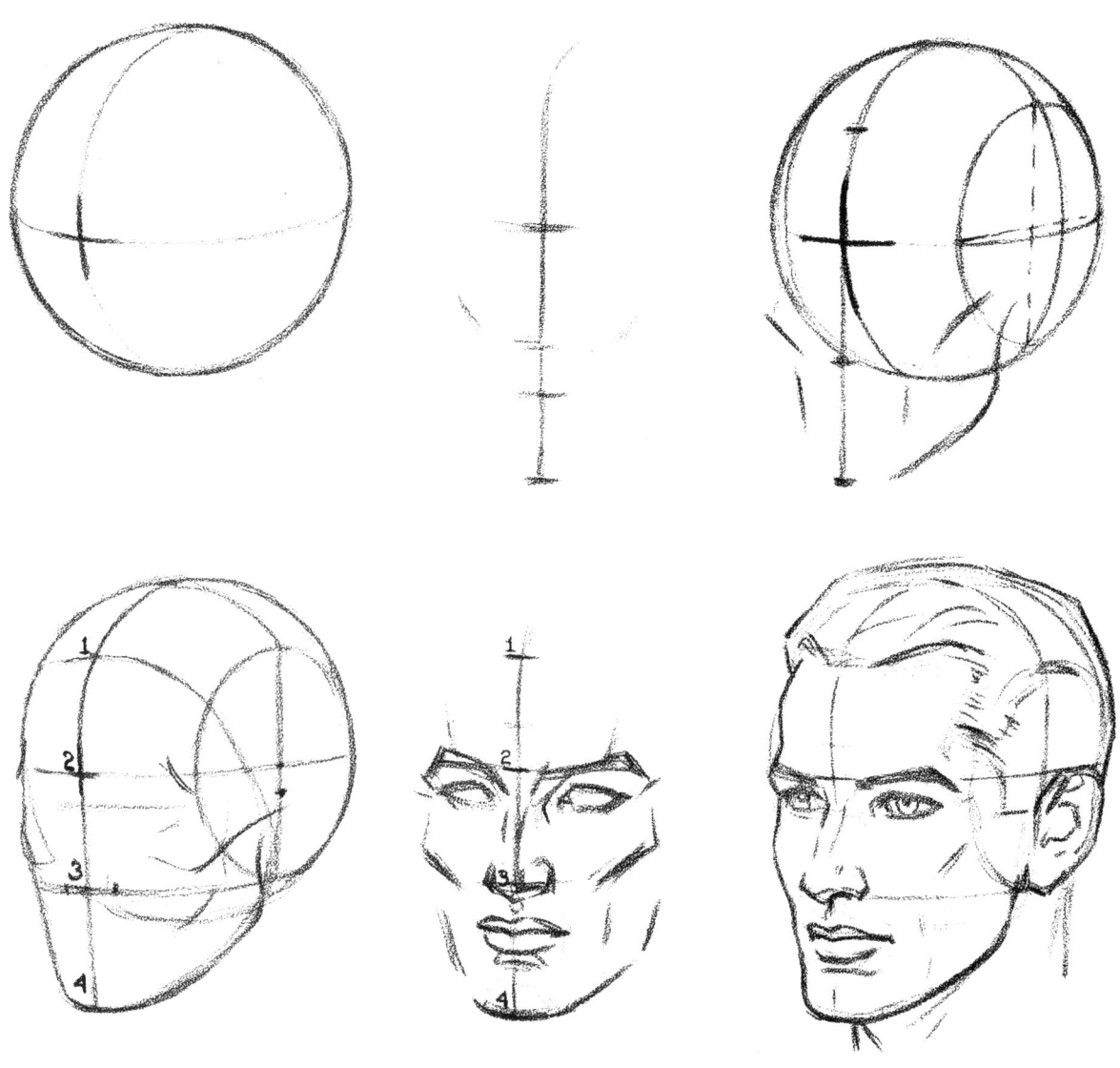

PLATE 4. Establishing the middle line

Start placing the features carefully.

If you have worked out the ball and plane and its divisions you will not have too much trouble in placing the features. However, you should realize that a feature will never fit on a head until it is placed correctly and in line with the construction lines of the whole head. Every artist must be prepared for a certain amount of struggle with construction, so do not allow yourself to get discouraged. Every head anyone draws depends on construction, just as much as every building, every car, every other three-dimensional object does. That is what the artist's job really is in learning how to construct things in three dimensions on a two-dimensional surface. We have to think of each thing we draw in its entirety and see how its dimensions appear to us from our particular viewpoint. Representation in three dimensions calls for knowledge and study. But such knowledge is no more difficult than that required for any other field. No matter how great your talent, talent has to work with knowledge to do anything well. When the search for particular knowledge becomes pleasant as well, half the battle is won. Construction need not worry you; it comes with practice.

26

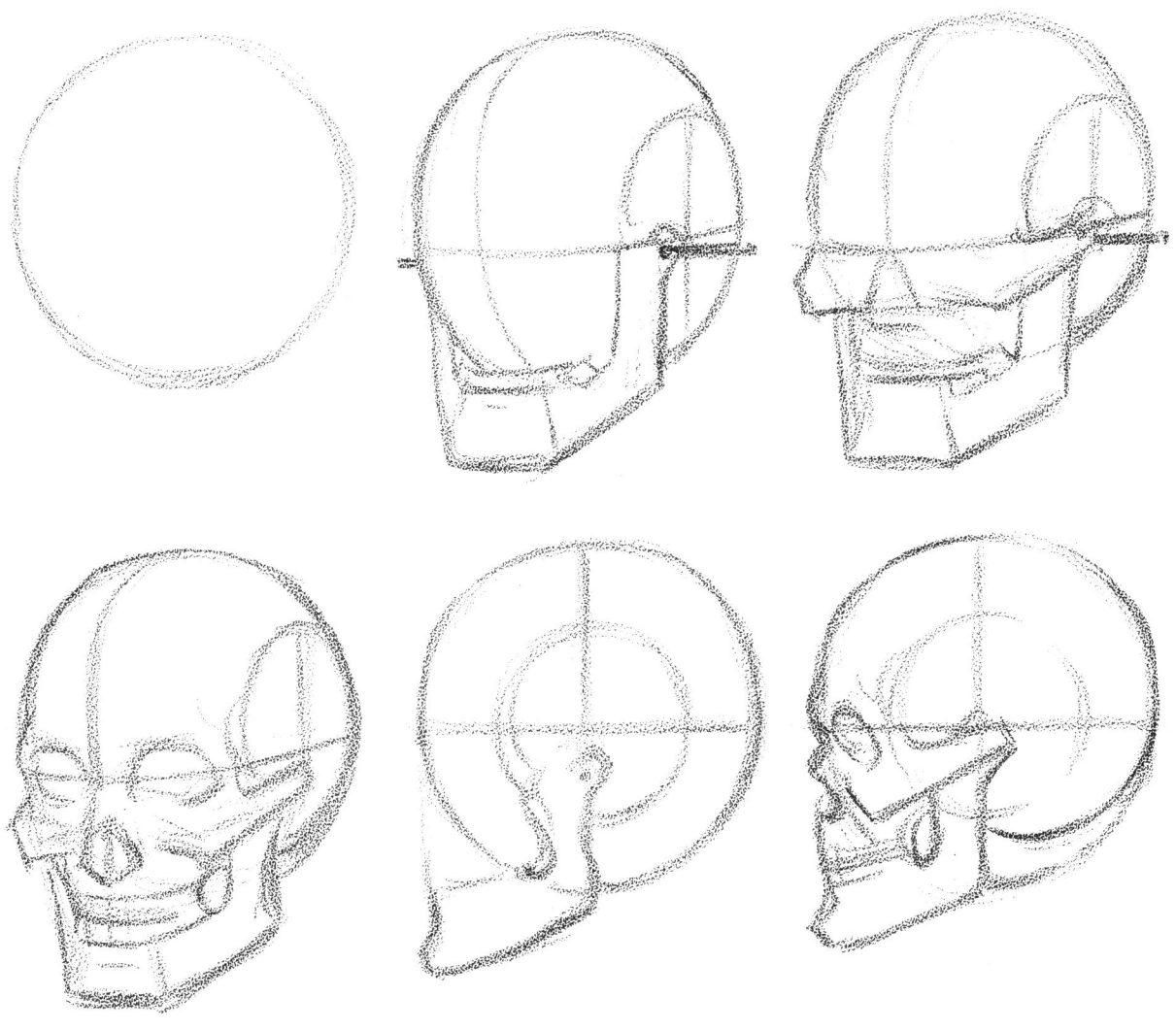

PLATE 5. Simplified bone structure

At this point it will help a great deal in constructing the head to have a fairly clear idea of the bone structure. Though we do not see the bones in detail, we must think of them as the framework of the head. All the division points of the head are related to the bones, not to the flesh. The reason we chose the ball and plane as an approach now becomes apparent, for our approach is the skull itself, simplified and made understandable.

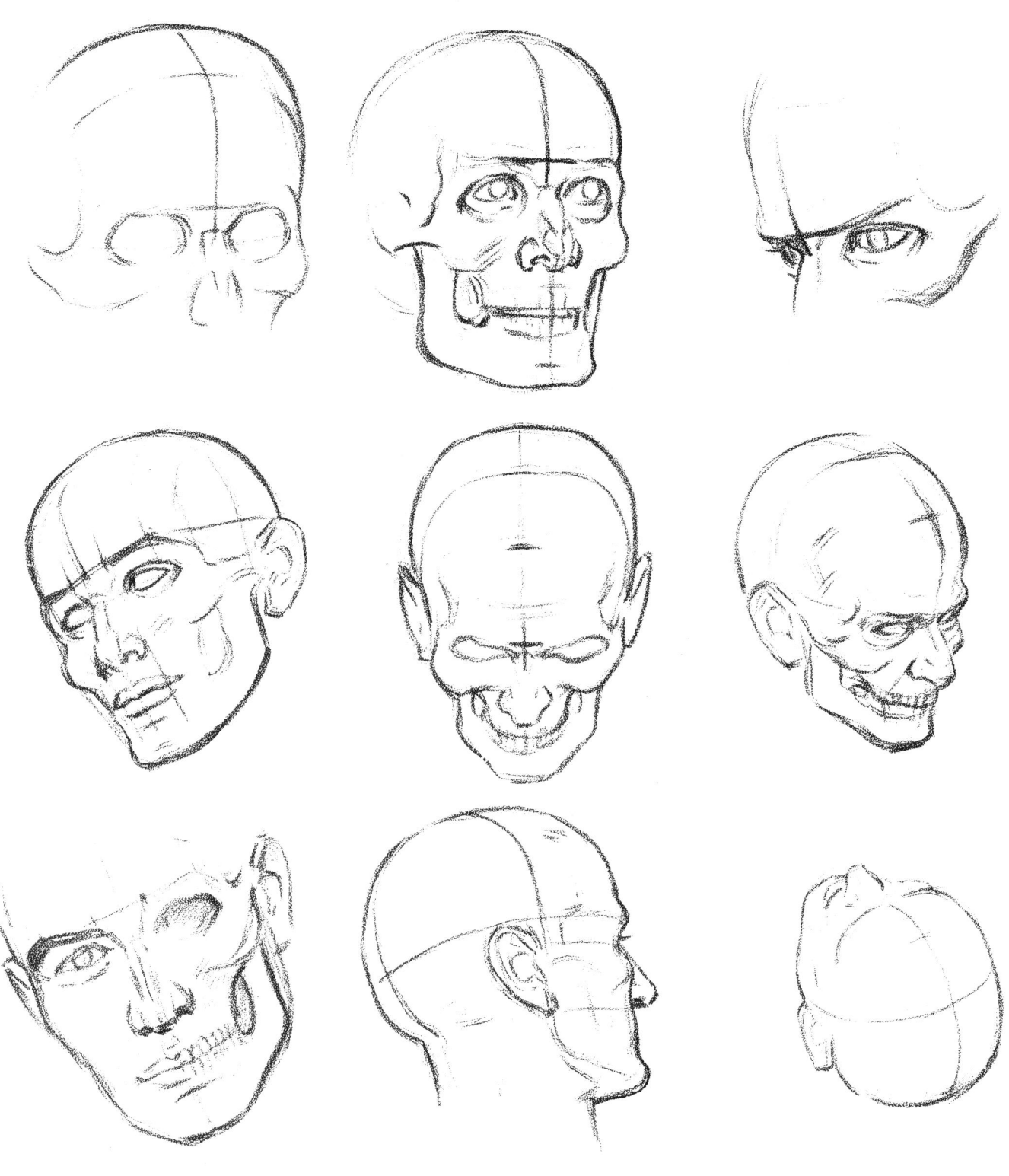

PLATE 6. The bony parts within the construction

Here we look at the bones more closely, realizing that, with the exception of the cheeks, all the flesh of the head lies over bone and is influenced by the shape of the bone. This simplifies our problem considerably, for except for the jaw the bones of the skull are all in a fixed position and move only as the whole head moves. Only the flesh around the eyes, the cheeks, and the mouth are capable of separate movement.

28

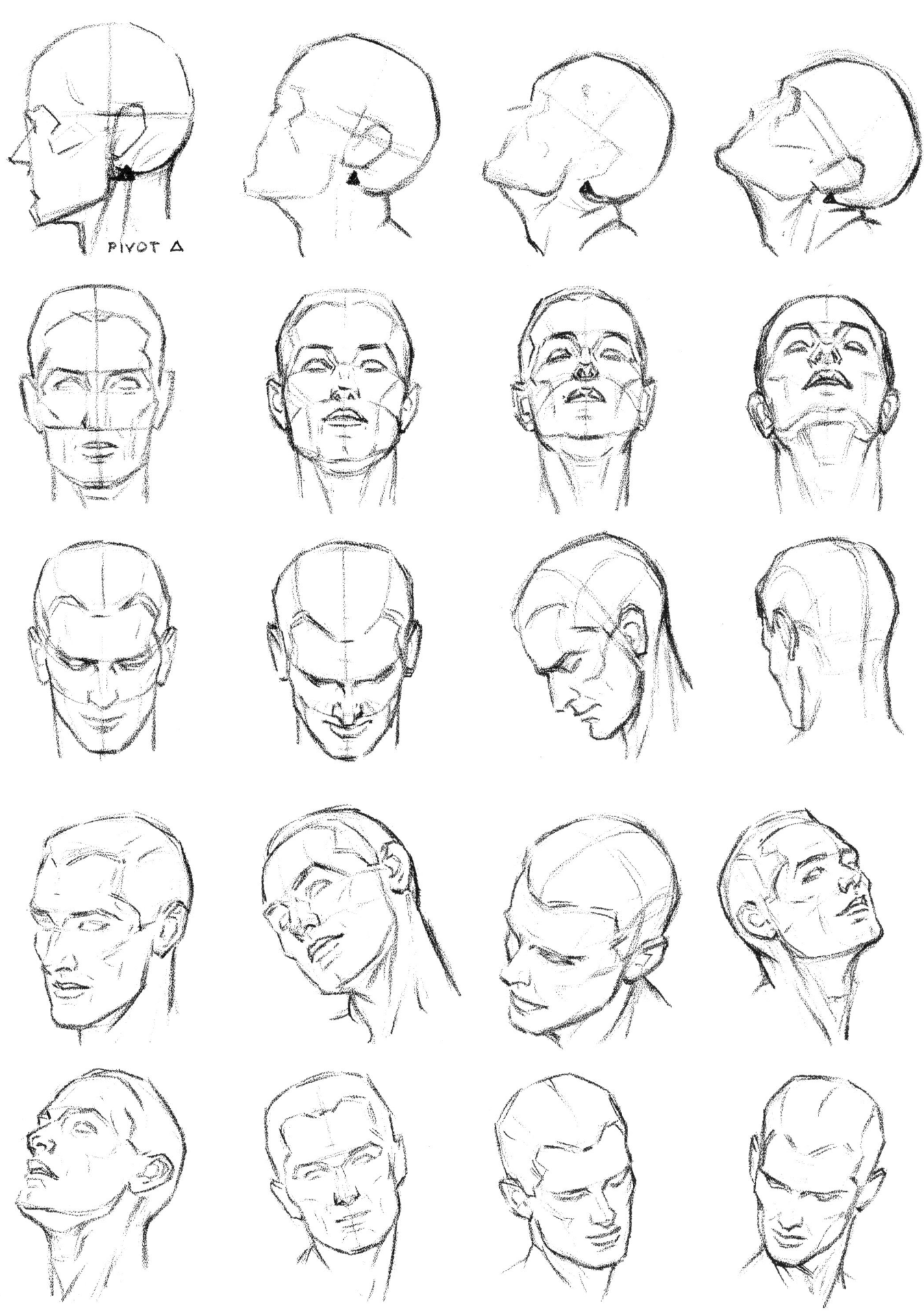

PIVOT △

PLATE 7. Action of the head on the neck

29

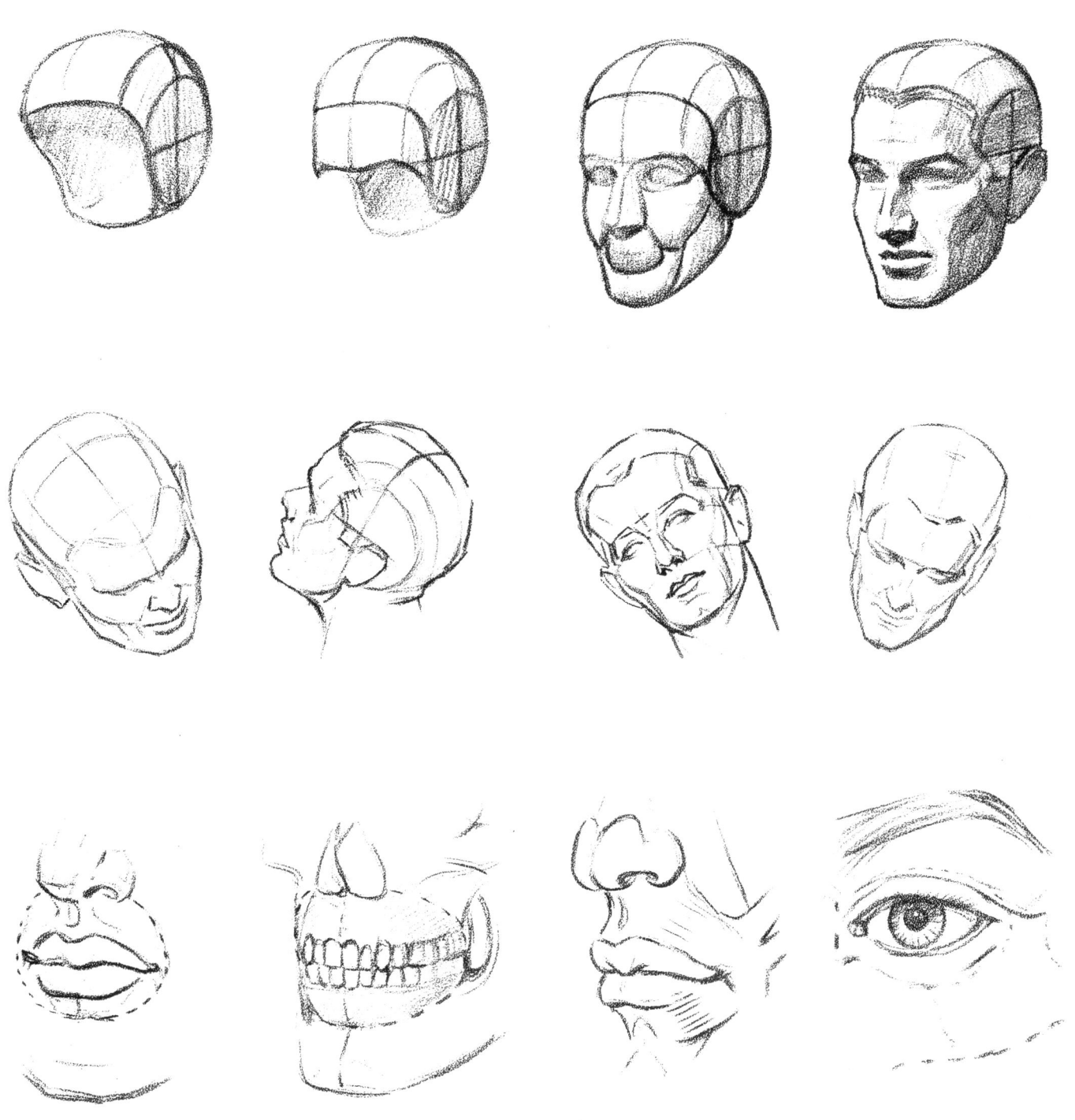

PLATE 8. Building the head out of pieces

If we think of the head as made up of separate pieces fitted together, we find the pieces shaped and put together as they appear in the drawings in the top row. Note the rounded piece which would contain the lips. We refer to this part of the skull as the "muzzle." In drawing the mouth we must make it fit around the curve of the upper and lower jaws and the front teeth. Too often the mouth is drawn as if it were flat against a flat surface. In the bottom row the three drawings at the left show the lips and the structure under them. The eye must also lie in its socket, as shown at the right. The eyelids operate much like the lips in closing over a rounded surface.

30

PLANES

We began by considering the head as round. This is logical, because it is much more round than square. However, one of the later discoveries in art was the fact that incessant roundness can become almost boring, and that a combination of roundness with squareness can produce a vigor of execution which many of the old masters lacked. The effect of roundness tends toward the "slickness" so frowned upon by modern artists and critics. Although the roundness exists, as photographs show, this type of rendition never seems to have the vigor of a drawing or painting in which the planes are stressed. For this reason a photograph of a head can never hope to compete with a good drawing as far as vitality of execution is concerned. It seems to me that the ideal lies somewhere between the two extremes. A drawing that is too square can look as if it were chiseled out of wood or stone, with more hardness than the subject warrants. On the other hand, a drawing that is too round may have so much sweetness and smoothness that it seems to have no structure at all beneath the surface; everything is polished and shiny. Of the two, I prefer too much character to too little. Artists have found that by squaring the planes, softening them only enough to relieve their broken-stone effect, they achieve solidity and vitality without going to extremes. It also has been discovered that flattened planes tend to merge into an effect of mere roundness at a distance. When you inspect a projection on a large screen from close up, it is surprising how flat the image is. However, if you step back, this flatness disappears and the full roundness seems to take over. The truth is that the halftones which model a surface are really much more delicate than they appear to be, and this truth has been a boon to painters.

For the time being, however, let us draw the planes as we feel they would really lie on the form. Through these planes we can interpret the true solidity as in no other way. It is better to learn to turn the form in its true structure than to omit the turning entirely so it may appear flat and without form. Remember that in a drawing the planes may be stressed considerably more than they can be in a painting, since we are dealing with fewer conflicting values. Just now we are not concerned with values, or "shading," as it is often called by the layman. We simply want to know what planes will give the basic form the general shape of the face and head. In other words, we want to get out of the round into more blocky forms, for this blockiness gives much more character, especially to men's heads. Turn to Plate 9. I suggest that you study this page carefully in order to fix these planes in your memory. They are like chords from which you build music; they are basic, and almost any head can be built on them.

After you have memorized these planes, try tilting the head and incorporating the visible planes, as shown in Plate 10. From these planes you can go on to perspective, as demonstrated in Plate 11. When you have mastered the construction of the ball and planes of the face, learned to use correct spacing and construction lines, and have assembled the planes, you have come a long way toward good drawing of heads. You should now be able to spot many of the difficulties that arise, and make the corrections in your basic drawing. Many a portrait has been started, only for the artist to discover after days of work that the basic construction is at fault. Something must be moved—an eye, the nose, or the mouth—and a likeness or the desired expression simply refuses to come about. A very good way of studying construction is to draw the construction lines on a clipping of someone else's picture of a head, so that you can see the exact placement of all parts. Once you understand the construction yourself, it becomes woefully apparent to you when the other fellow does not. Some very clever artists do not really know how to construct correctly, and they spend many hours of added difficulty as a result. No "knack" of drawing heads can compete with sound knowledge.

In Plates 12 to 16, I have planned a little fun for you. We start taking some liberties with the basic ball and planes. You will do this better without copy. We do some experimenting with types, as I promised early in the book. To produce different types we can vary the ideal or average measurements. The three divisions of the middle line of the face can be made unequal, or exaggerated as you wish. Then we can vary the shape of the cranium and bony understructure. I suggest that you play with expressions and characterizations. It is interesting and sometimes amazing what you can produce in the way of characters by variation in the spacing and basic shapes. You hardly know before finishing what type you will end up with. On the other hand, you can actually plan a given type and come very close to achieving the result you want. You will find yourself drawing heads that are most convincing, that have even a professional look. I suggest you try beards, mustaches, high or low, thin or heavy eyebrows, big noses, little noses, jutting chins, receding chins, narrow heads, wide heads, flaring jaws, and what not. Have some real fun while you are at it. You may or may not be interested in cartooning, but it is fun to draw characters, and you will find that you can do better than you might have thought possible. Watch the perspective and construction as carefully as you would in drawing any head, but exaggerate all you can. A good way to experiment is to jot down beforehand a little description of the character you wish to draw, then try to draw the head you have described. Next, ask someone else to give you a description of a character. Try that. Such practice means that you can, at an early stage of your knowledge, begin to create, as you would if you were an illustrator. Stick fairly close to outline heads just now, but try to create the type you want.

As an example, your description might be something like this: "John is big and raw-boned. His eyes are deepset under shaggy brows. There are hollows under his cheekbones. He has a big nose, heavy jaw and chin. His hair, though thin on top, is bushy around his ears and the back of his head. His eyes are small, dark, and beady." Now try to draw John with the knowledge at your present command.

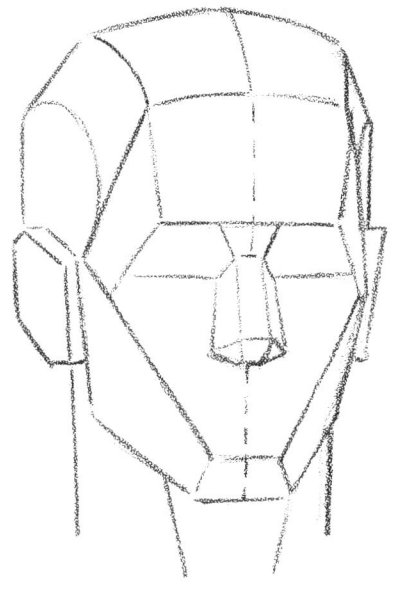
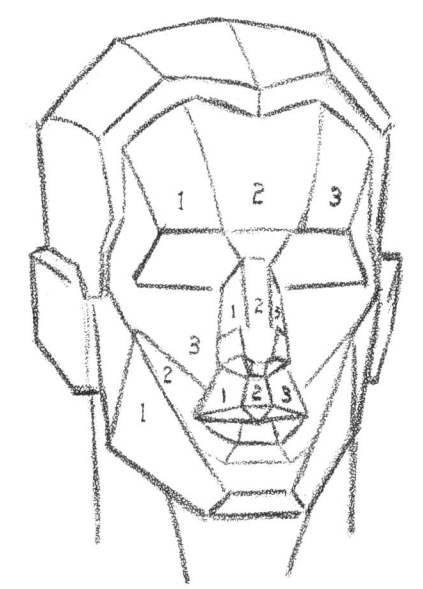

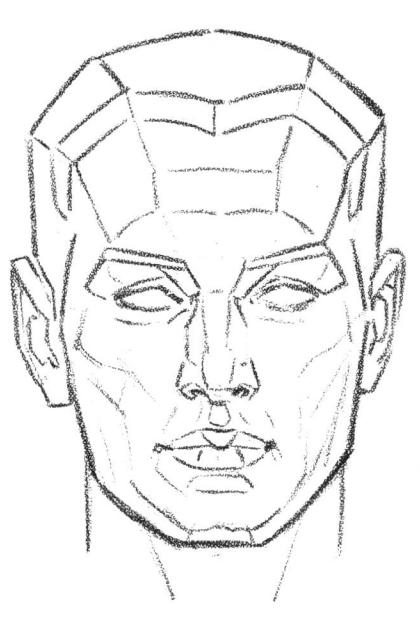
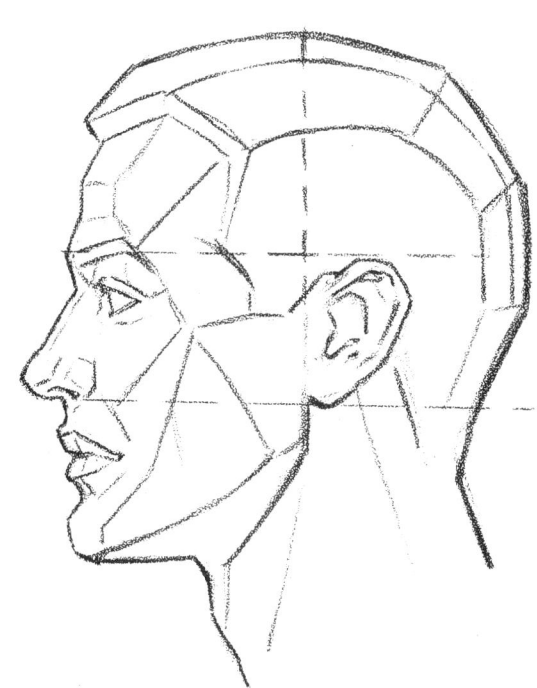

PLATE 9. Basic and secondary planes of the head

The planes of the head should be memorized, for through them we have a foundation for rendering the head in light and shadow. Begin with the basic planes (top, left), and study them until they are fixed in your mind. Then take up the secondary planes. From these sets of planes almost any head can be built. The surface varies with the individual character, but with the planes shown here you can produce a well-proportioned, manly head.

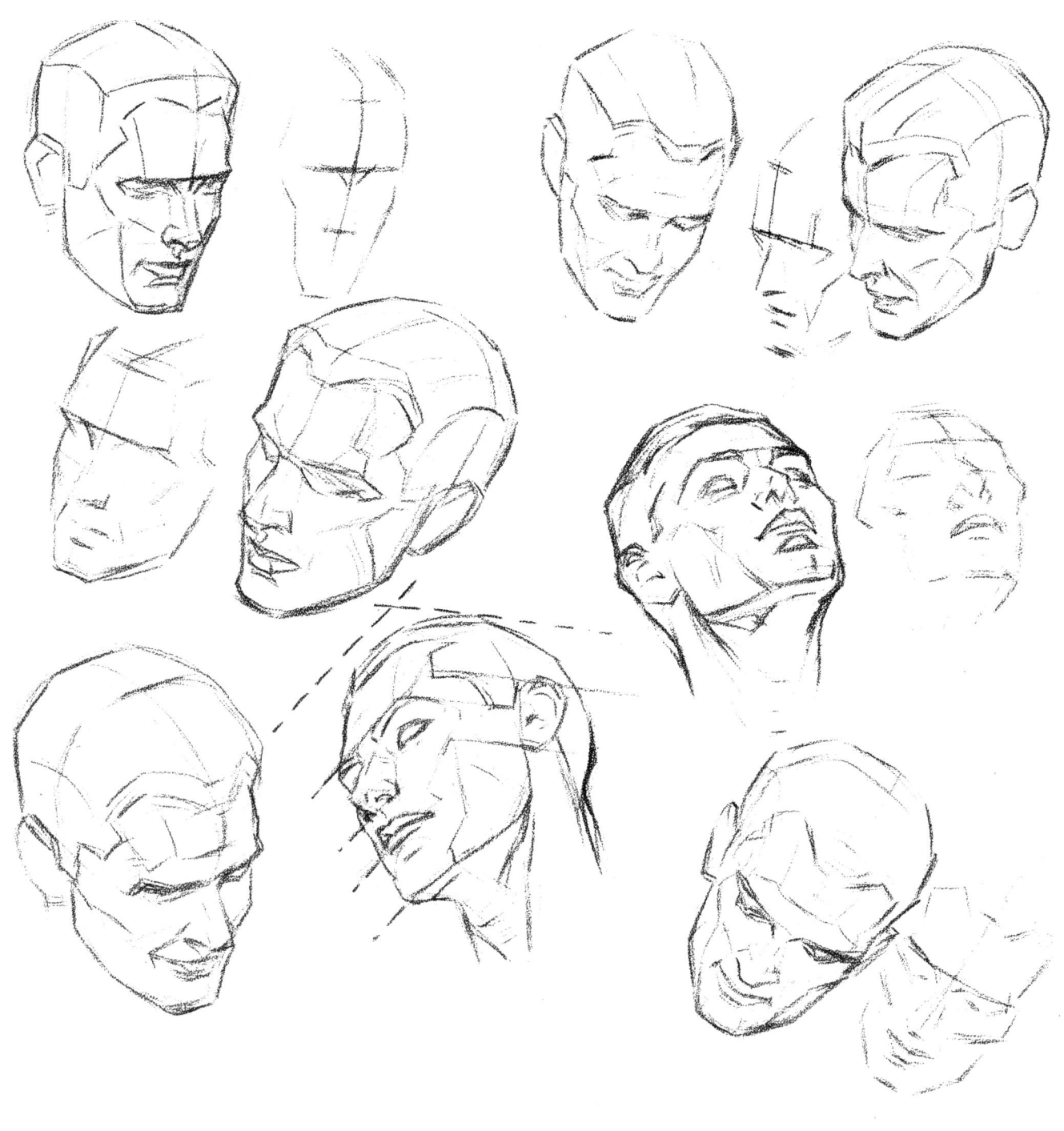

PLATE 10. Tilting the head

Planes help us to maintain construction throughout the face and head, within the construction lines or divisions of the basic ball and plane. The muzzle becomes easier to draw in all sorts of tilted positions. The slant of the cheeks and the rounded rectangle of the forehead fall into place within the three divisions of the face. By thus representing the head in block form, we determine the angles throughout the head. This is our first step toward the perspective of the head.

34

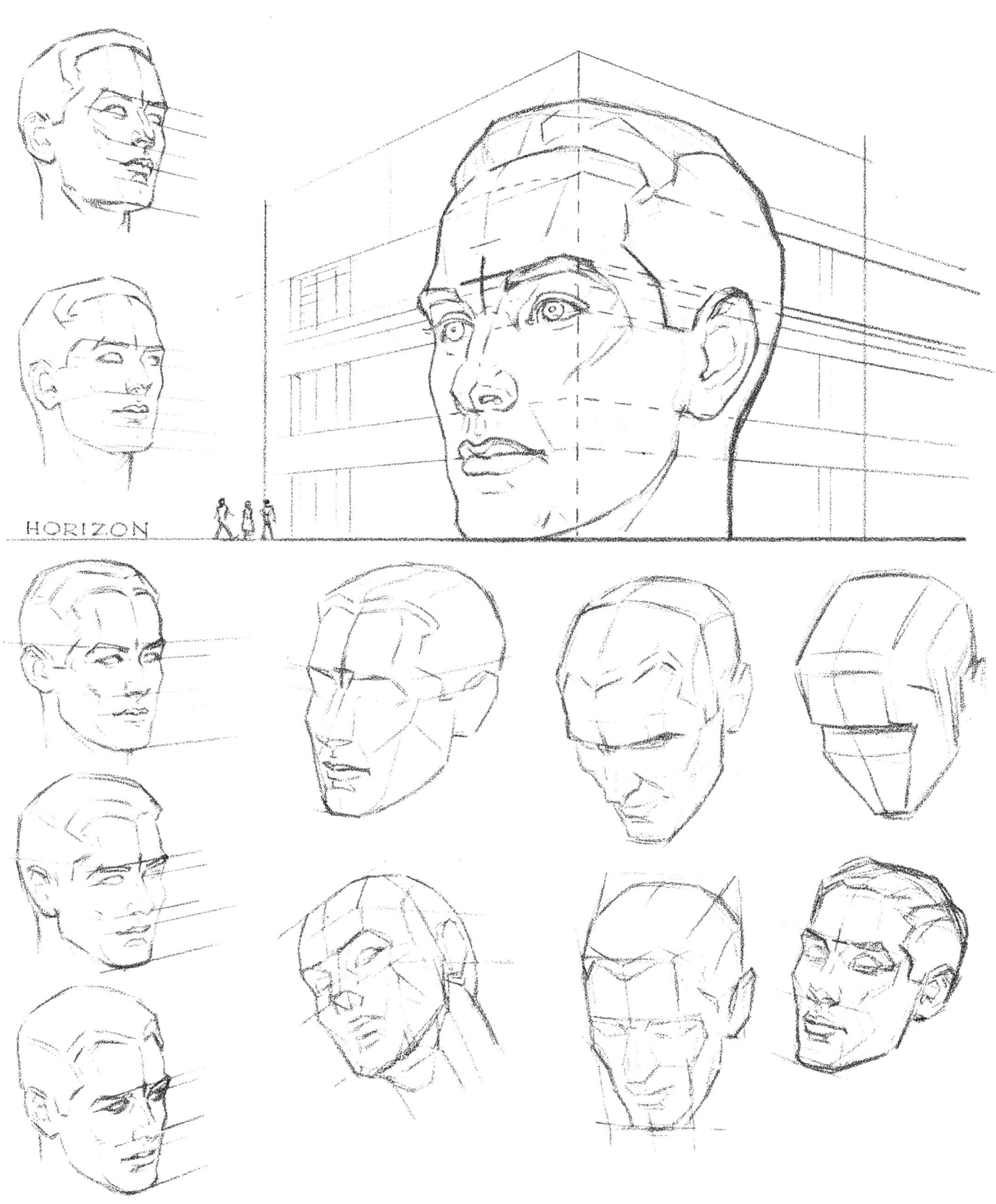

PLATE 11. Perspective in drawing the head

The handling of perspective marks the difference between the amateur and the professional. Every object drawn has to have an eye level or horizon, felt if not actually represented. On the left we see the planes of the head as seen from above or below the eye level. If a head were as big as a building it would be affected by perspective in the same way as a building is.

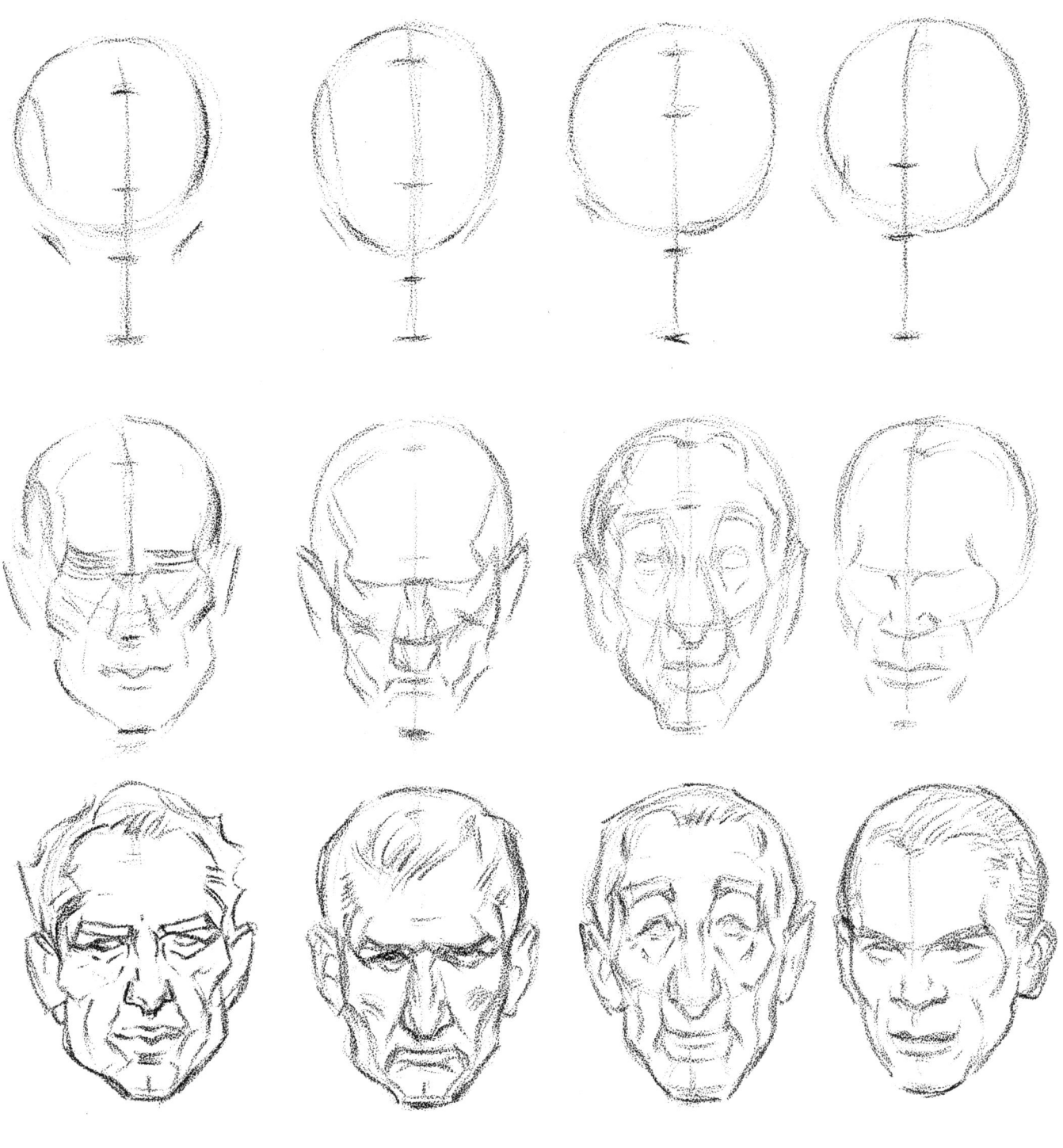

PLATE 12. Variety in spacing creates types

In order to create differences in type and character, we may decide not to follow the basic measurements or divisions too meticulously. By varying the proportions of the three divisions of the face, we come up with a good deal of variety in the results. There are thousands of possible combinations. It is fun to experiment with them.

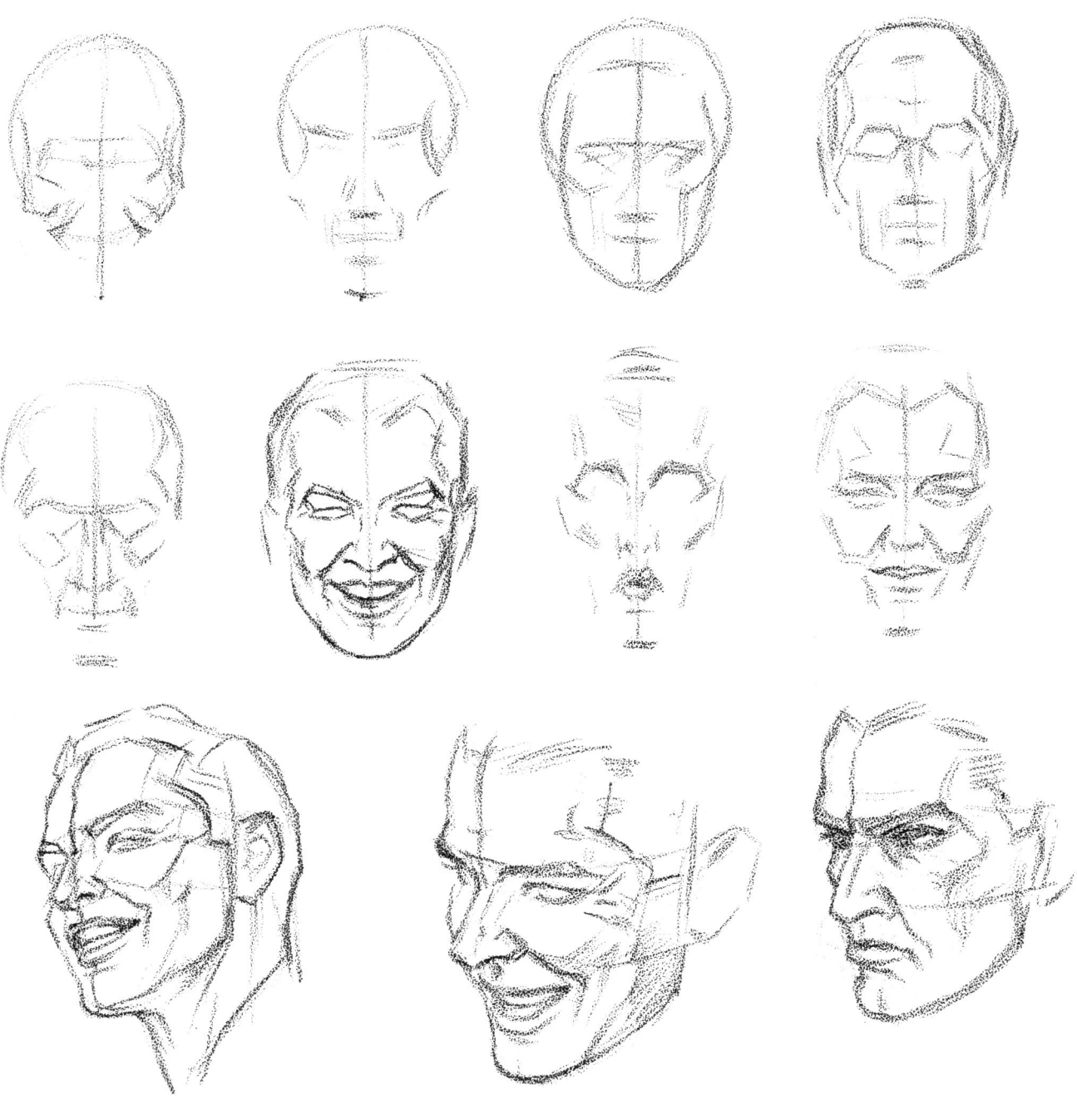

PLATE 13. Always build on the middle line

Always remember when drawing a head to balance the forms on both sides of the middle line. The bony parts stay fixed, and the expression fits in between. All the jaw can do is open and close. The expression lies in the eyes, cheeks, and mouth, with some wrinkling of the forehead and around the eyes. What we do on one side, we must do on the other.

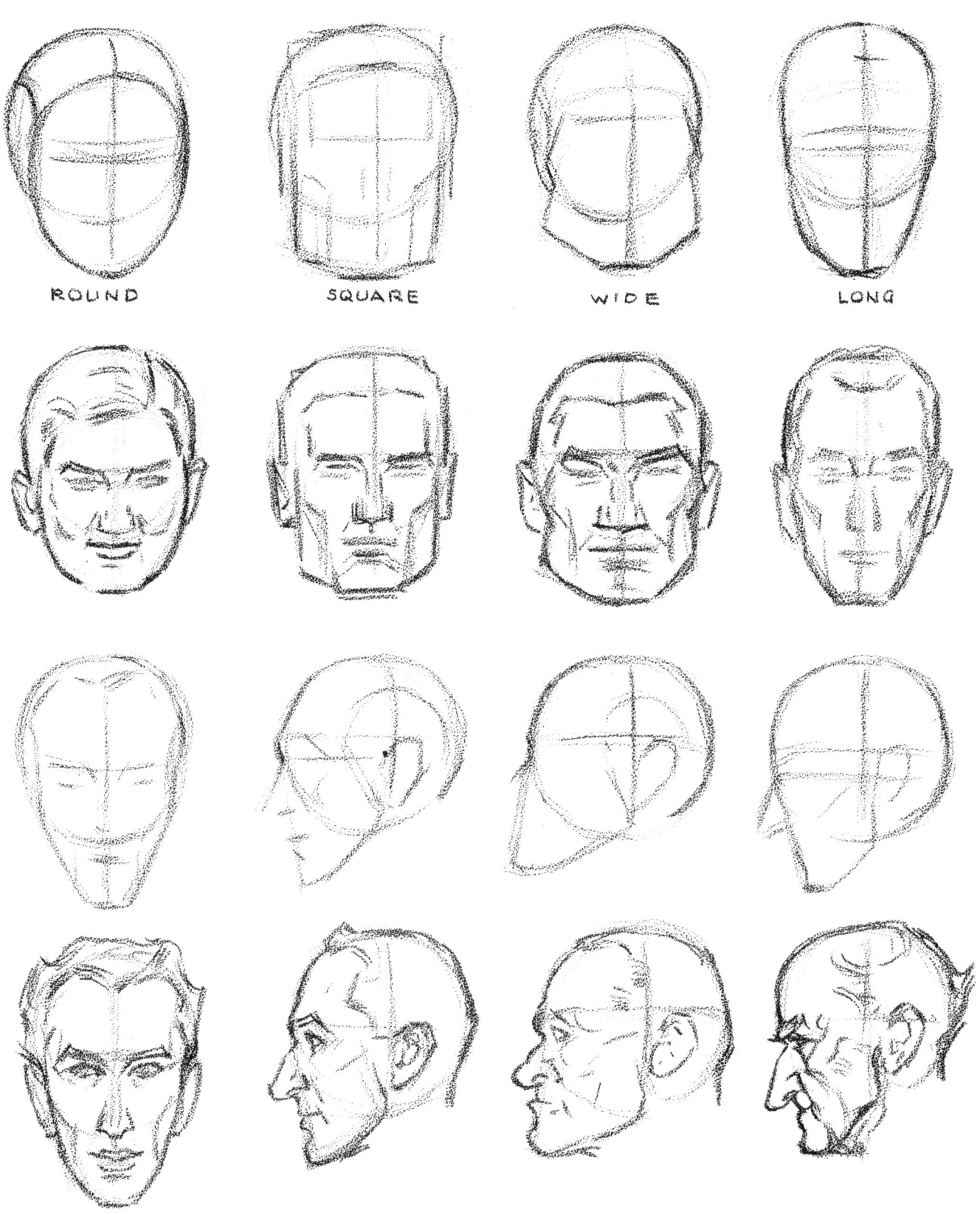

ROUND   SQUARE   WIDE   LONG

PLATE 14. Creating any desired type

There is no reason why you can't take all the liberties you wish with the ball and plane. The variety of types mentioned in the early part of the book are drawn simply by building an understructure that is wide, square, long, narrow, or anything you wish. You have the basis of construction, so now just try some variations.

38

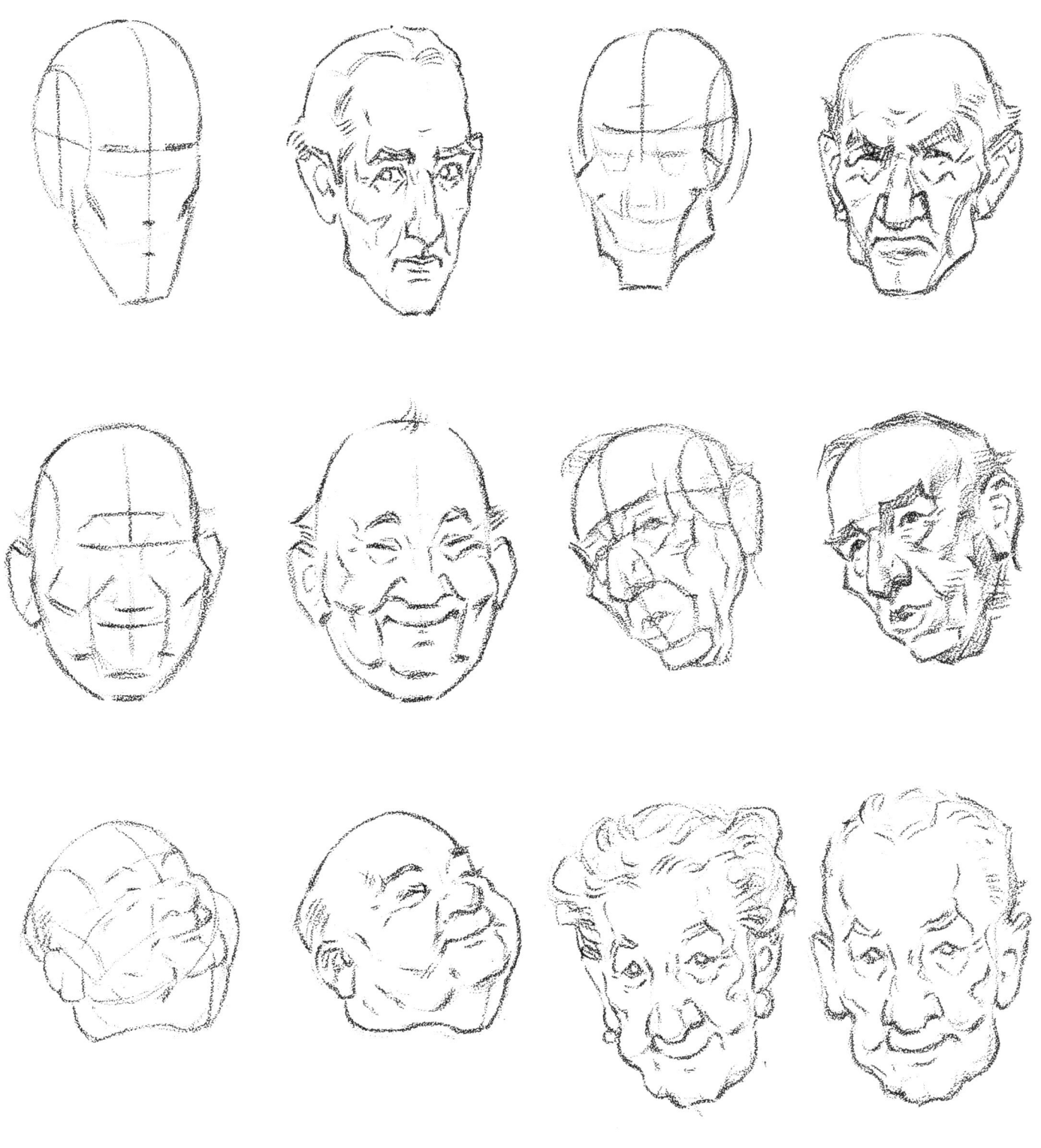

PLATE 15. Types are built by varying the ball and the plane

Look about among the people you know and those you see around you. Study them with a new understanding. See the combinations created by nature. Look from hairline to brow, then at the middle area from brow to bottom of nose, and finally to the bottom of the chin. Look down the middle line of a face; study what you see on each side.

39

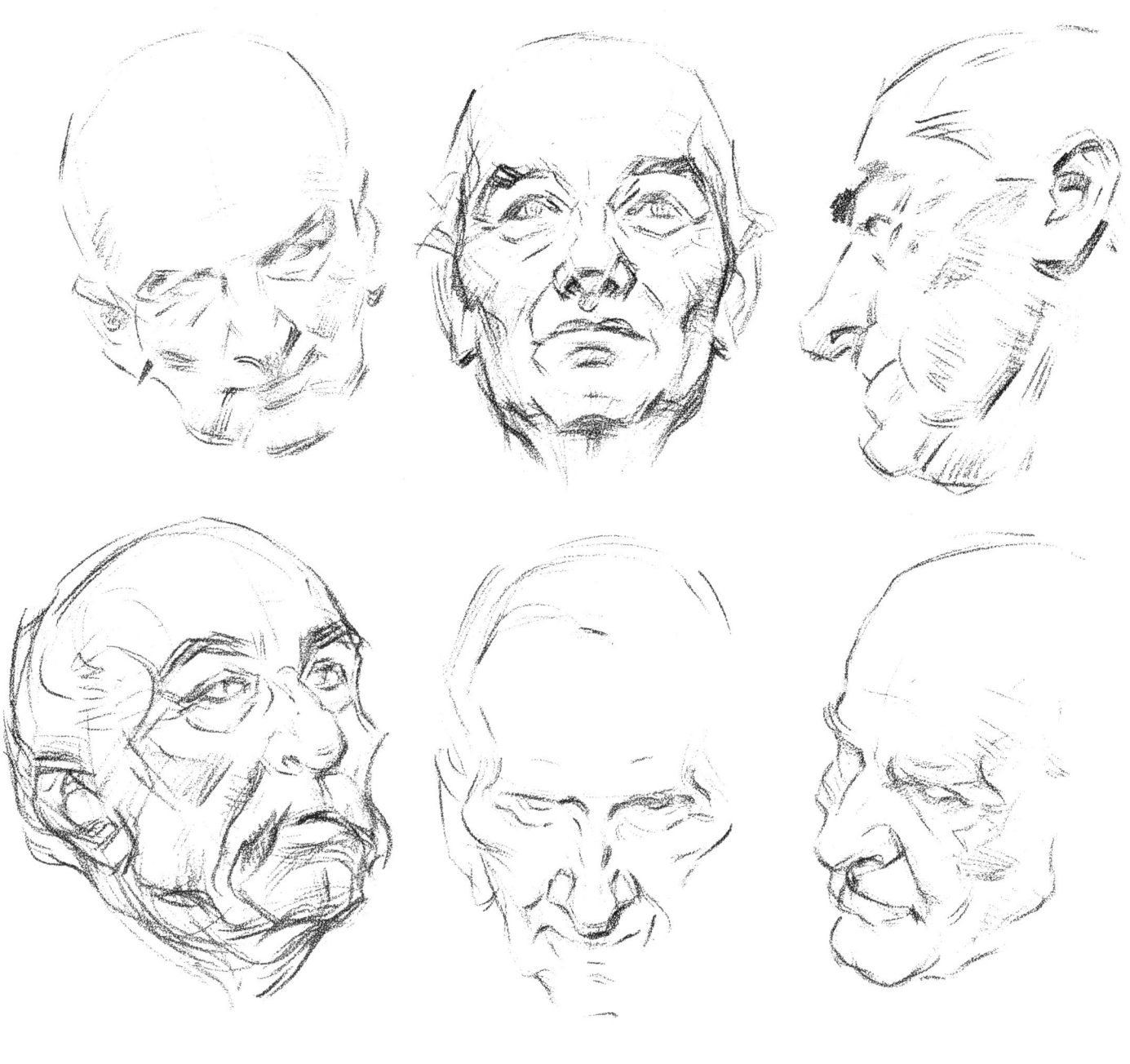

PLATE 16. Indicating character

Once you know how the lines of construction are set up in a head, you can quickly analyze faces and skulls. Always look first for the bony shapes, and the location of the features. Then look for the flesh formations in the cheeks, around the mouth, and around the eyes. Such formations can be easily indicated. See if the cheekbones are prominent and accented by shadow shapes under them. Look at the nose and the formation of the nostrils, the lips, and the creases between the lips and cheeks. Follow the shapes down into the chin and along the jaw line. These general characteristics, along with the whole shape of the head, are more important than a photographic delineation of each square inch of surface. Older people are more interesting than the young for this sort of study, since the characteristics have had a chance to develop.

RHYTHM

Rhythm in drawing is something you feel. Rhythm must be closely associated with design, and every head has design. There is a related flow of line, one line working with or opposed to another. Rhythm is freedom in drawing, freedom to express shapes, not meticulously, but in harmony. Rhythm is the hand working with the brain more than with the eye, the feel of the thing rather than the look of it. In drawing, rhythm comes with practice just as it does with a golf club. No one can tell you how to acquire it, but as you become conscious of it, you begin to recognize it when it is there.

To try to describe rhythm in drawing let us say that the artist is feeling the simplified shape of the whole thing as he draws every part of it. You see his hands swinging over the paper before the pencil goes down. He feels the stroke before he makes it. Rhythm need not always be curves. Curves may oppose blockiness. Rhythm might be an accent where it will do most good. It is more often the suggestion of the form rather than the closely scrutinized detail of the form. Here again the artist leaves the camera far behind, for the camera must record detailed fact, and only when rhythm is set up before it can it catch this elusive quality. The onlooker senses rhythm in your work even if he cannot consciously define it. You sense rhythm in some handwriting, while other specimens are cramped, jerky, and scrawly.

Some people have natural rhythm; others must strive to acquire it. Take the pencil in the palm of your hand between the thumb and first finger rather than holding it as you would to write between tight, cramped fingers. Swing it over your paper, using your wrist and arm and keeping your fingers still. That is the way to draw a rhythmic line. You can train your hand to draw, instead of using the fingers. Movement becomes associated with the whole arm rather than with the fingertips. Draw things large for a while. George Brigman, the famous anatomy teacher, used to illustrate his lectures by drawing with a crayon on the end of a four-foot stick. Some of his anatomy drawings were many times larger than life, and they were beautiful.

Rhythm is all about us, but we must train ourselves to see and recognize it. It might be described as the longest line, straight or curved, that you can make before the direction of the edge changes. A long direct line is more expressive than a myriad of little whiskery lines. An arrow in flight is a perfect example of rhythm. The movement of water or waves is another. The arc of a baseball in the air, the way a fielder drops his hands in the line of flight as he catches the ball, the movement of the forms in a woman's hair—all have rhythm. We might call it the uninterrupted flow of line which seems to reflect the movement of the artist's hand.

I cannot tell you how to acquire it, but I do believe you can. Awkwardness comes from lack of training; rhythm from trained organization, or coordination, perhaps both—knowledge and ability working together. Rhythm is one thing no camera or projector can ever give you. You feel it and strive to express it, or you don't. Swing that pencil over your paper just to draw a free line. Nobody ever does it too well the first time he tries.

41

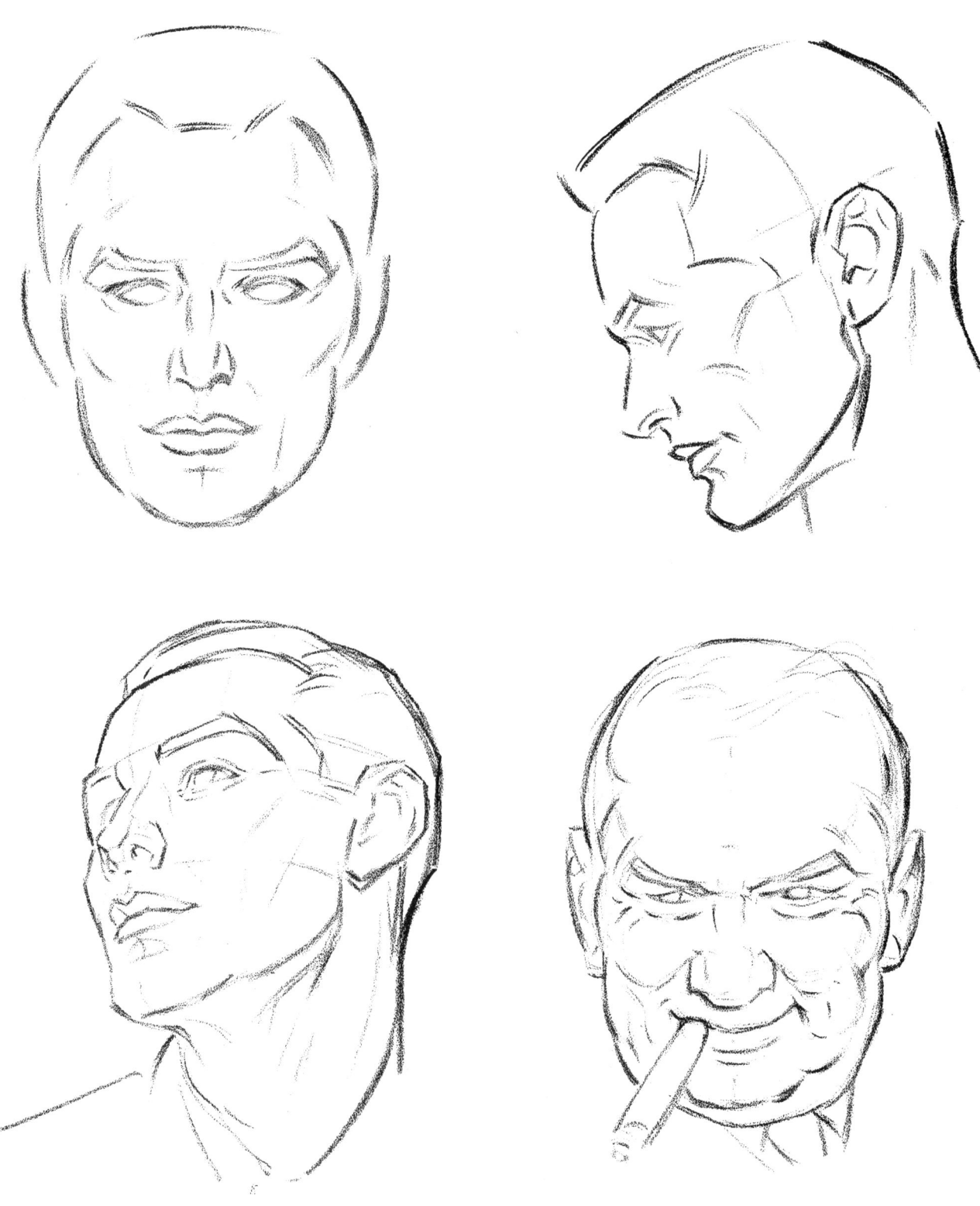

PLATE 17. Rhythmic lines in the head

It is interesting to search for the rhythmic lines in faces. You will find rounded or curved lines in opposition to angular and blocky lines. The blocky treatment helps to get away from the tight photographic approach. Then the head looks drawn, not traced. There is charm in curves but square forms have weight and solidity. You can produce happy results by combining the two instead of merely copying every waver of every edge in exact outline. In this way you set a feeling of design, and at the same time render solid form.

THE STANDARD HEAD

Heads will naturally vary in measurement and proportion. However, any artist will find it most practical to carry in his mind as basic measurements a scale of proportions, built on averages and simplified. The front view of the head fits quite well into a rectangle that is three units of measurement wide, and three and a half deep. This scale leaves a little space beyond the ears on each side. The half measurements of these units locate the eyes and nose and help in placing the mouth, and also put the line of the eyes at the halfway division of the whole head from top to bottom, as it should be and as

it averages out in a large percentage of actual faces. This method of unit measurement locates the hairline and the three front divisions of the face. The side view of the head fits exactly into a square three and one-half units in each direction. You can establish your own unit; it is the proportions that are important.

These proportions, shown in Plate 18, have been worked out after a great deal of research and are offered to meet the need for a simple and practical scale that is readily usable. This scale fits perfectly with the ball-and-plane approach.

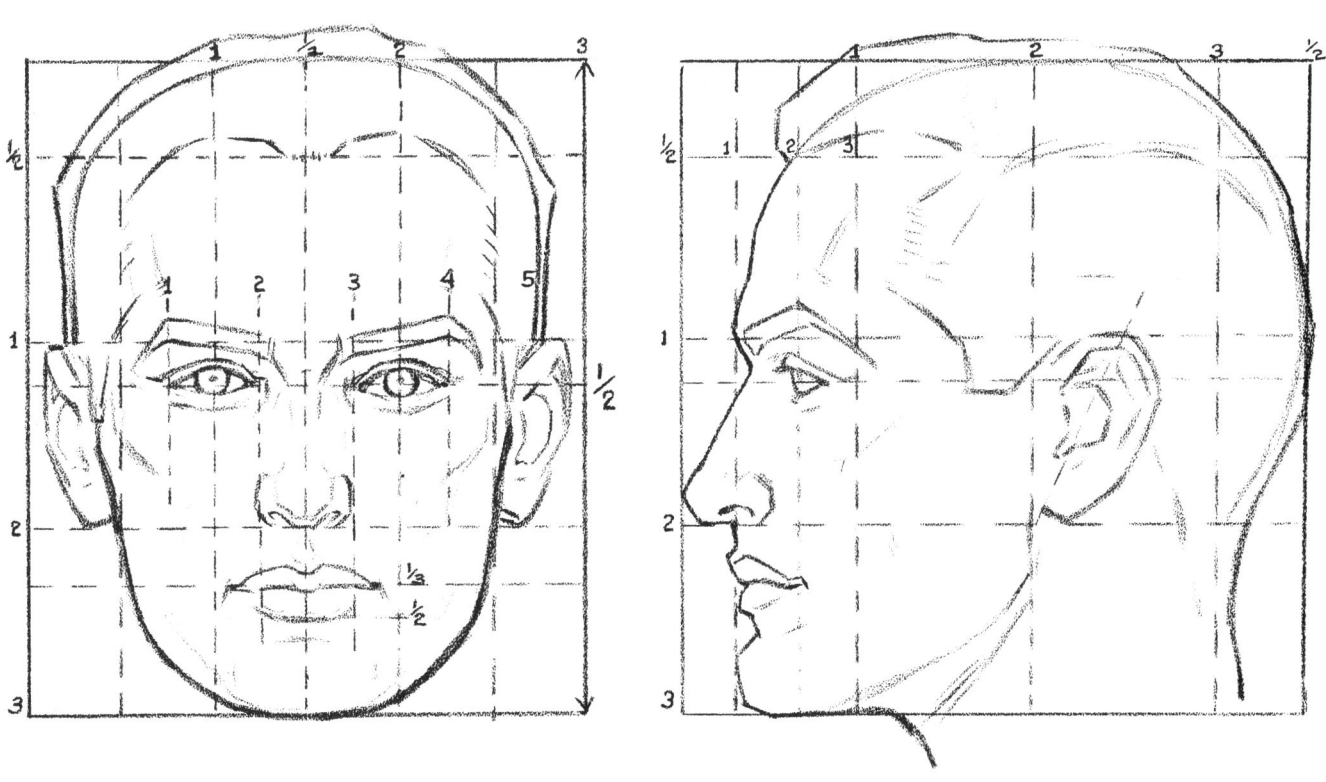

PLATE 18. Proportions of the male head

The standard proportions for a man's head are worked out here for the front view and the side view. The scale may easily be memorized. The head is three and one-half (optional) units high, nearly three units wide (to include the ears), and three and one-half units from tip of nose to the back of the head. The three units divide the face into forehead, nose, and jaw. Ears, nose to brow, lips and chin are each one unit. So you may start in this way to draw a head in any size you wish, using your own unit of measurement.

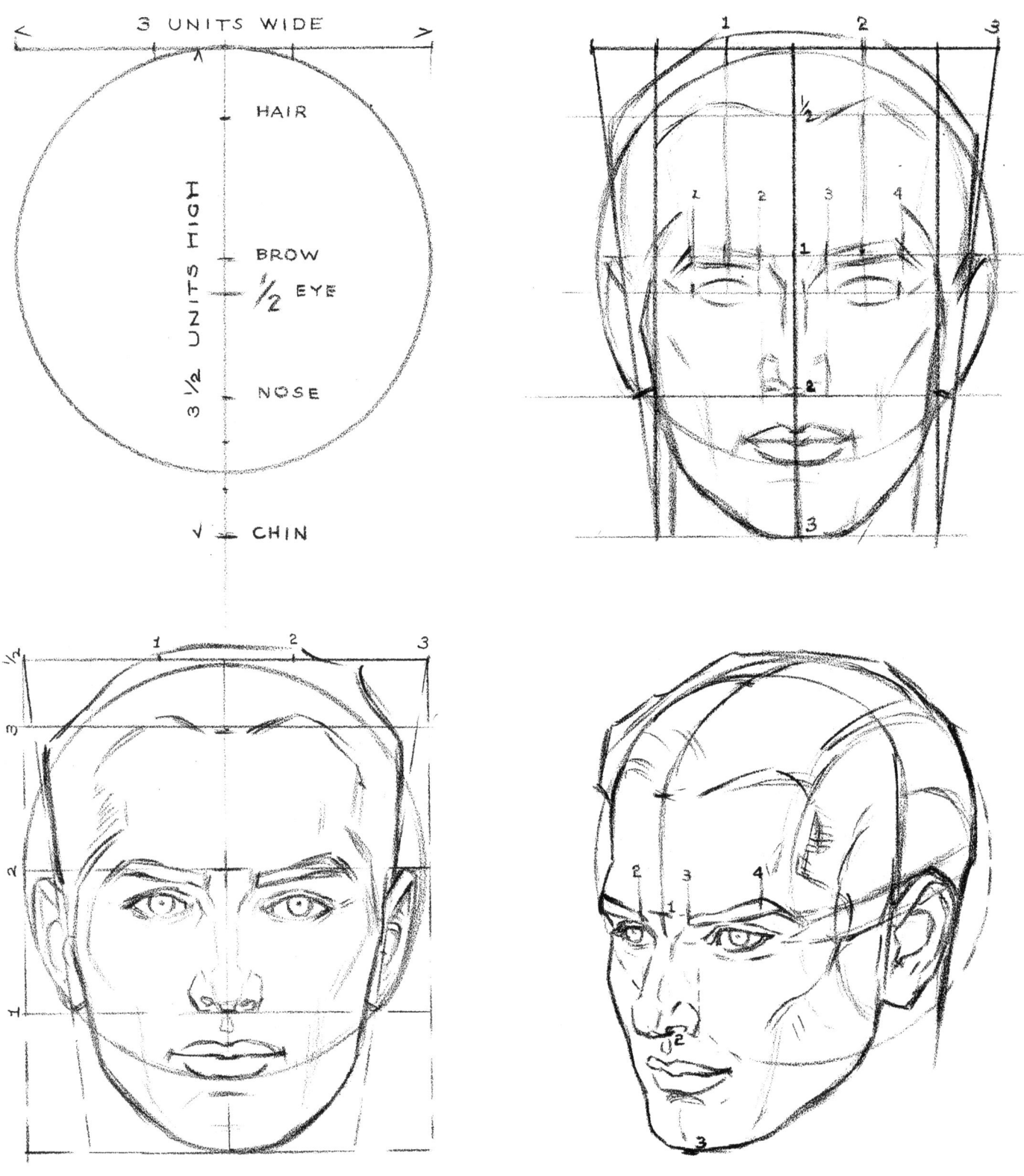

PLATE 19. Drawing the head in units

Here you see how the scale works out in practice. The circle represents the ball, and the width is the width of the head, including the ears. We find that the face is about two units wide and that the eyes fall between the middle halves or at the quarter points of the two units (see upper right). This coincides with the divisions of the ball and plane with which you are already familiar.

44

MUSCLES OF THE HEAD AND FACE

I do not see any material advantage to the artist in knowing the names of all the muscles and bones of the head, but it is of great importance to him to know where they are, where they attach, and what they do. It is important to know that some muscles are attached directly to bone at both ends, while others are attached to bone at one end and to other bands of muscles at the other. The former have the function of moving the bony structure. The latter move the flesh. Plate 20 shows the muscles and how they are connected.

The most important muscle of the head is the powerful muscle that closes the jaw. You feel it at the corner of the jaw, just below and in front of the ear. Circus acrobats have been known to dangle the weight of the whole body at the end of a rope by biting a bit of hard rubber attached to the rope end. The jaw is also attached to a muscle that spreads out over the side of the cranium. These two muscles give the power to crunch and grind food in the mouth.

A very marvelous mechanical principle functions in the eyes and mouth. Both are slits in a circular sheet of muscle. If you took half of a hollow rubber ball and cut a slit in it, without stress on the rubber, the slit would close itself. Under tension you could easily pull the slit open. The dropping of the weight of the jaw opens the mouth. To open the mouth wide is a conscious effort. To keep the mouth closed really requires very little effort—a piece of knowledge that can be used to great advantage at times.

Very important are the little ribbon-like muscles which open the lips laterally, pulling at the corners of the mouth. These are the "smile muscles." They are the ones that puff the cheeks by contracting within the flesh. When they pull diagonally upward and a smile flashes, great things may happen, far beyond mere mechanics. Remember these as the "happy muscles." They attach at the cheekbones and run diago-

nally down the cheeks to the muscles around the lips.

Note the muscles which run down the side of the nose past the corners of the mouth to the chin. These are the "unhappy muscles." Being attached to the bone around the nose at one end and to the jaw at the other, they can pull the lips upward in a snarl or downward in a leer. Working from both ends, they expose the teeth the way an animal shows its fangs. These muscles are operating from both ends when you brush your teeth. They seem to pull downward when you are lifting a heavy weight, or in extreme muscular effort of the body, like running. They make round corners at the mouth, where in the smile the corners are pulled out and upward. Try to associate the happy and the unhappy muscles, for they are the basis of most facial expressions. The wrinkles at the corners of the eyes are simply caused by the flesh of the cheeks' buckling by the upward pull of the "happy muscles" below the cheekbones. The bulging of the cheeks also causes the crease or fold of flesh under the eyes in a smile. It is more pronounced in some faces than others. As the "happy muscles" pull at each side in the smile, the nostrils may flare a little and become more evident, which is one of the things that help to make a face smile.

The dimple or downward line occurring in the lower part of the smiling cheek is caused by the little open space between the "unhappy muscle" and the jaw muscle. In old age this depression becomes very evident. In the young face it is a dimple.

The rest of the face muscles are simply what we may call "wrinkle muscles." There's one at the inside corner of the brows near the nose. This one lifts the corner of the eyebrow as in worry or in an expression of pleading. The "unhappy muscle" pulls down the inside corner of the brow in a frown. The two "wrinkle muscles" above the brows also wrinkle the forehead, since they are contracting beneath the flesh, but are also attached to the flesh.

There are two small "wrinkle muscles" at the point of the chin. The depression between these muscles may account for a dimple in the middle of the chin. They also cause the chin to buckle into little bumps in some expressions.

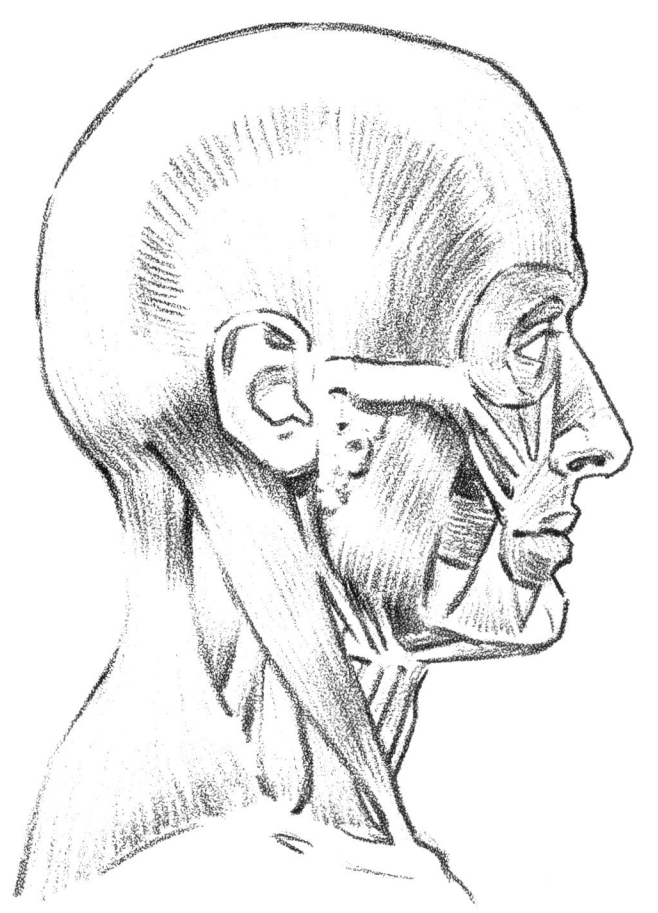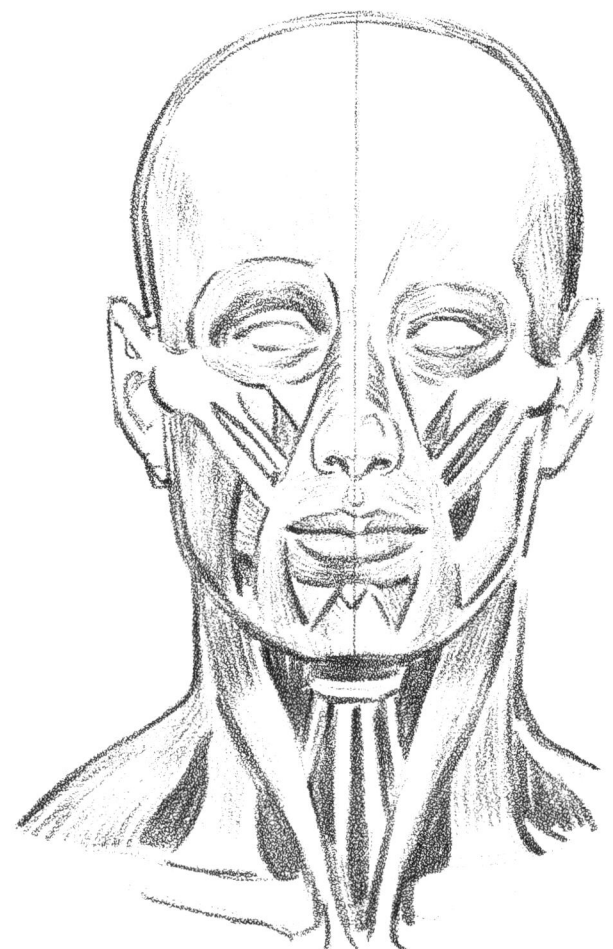

PLATE 20. Anatomy of the head

When you are studying the muscles of the face, get in front of a mirror and give them a good working over. From that and from these drawings you will learn a great deal about expression and the why of it.

Give some consideration to the muscles of the neck, for you usually have to draw a head on a neck. The two diagonally placed muscles that turn the head are attached to the skull just behind the ears at the top, and to the breastbone, which lies between the two collarbones, at the bottom. Two strong muscles attached to the back of the head underneath the back of the skull hold the head up or tip it backward. The head drops forward mostly of its own weight.

To know these muscles will help you tremendously in drawing heads.

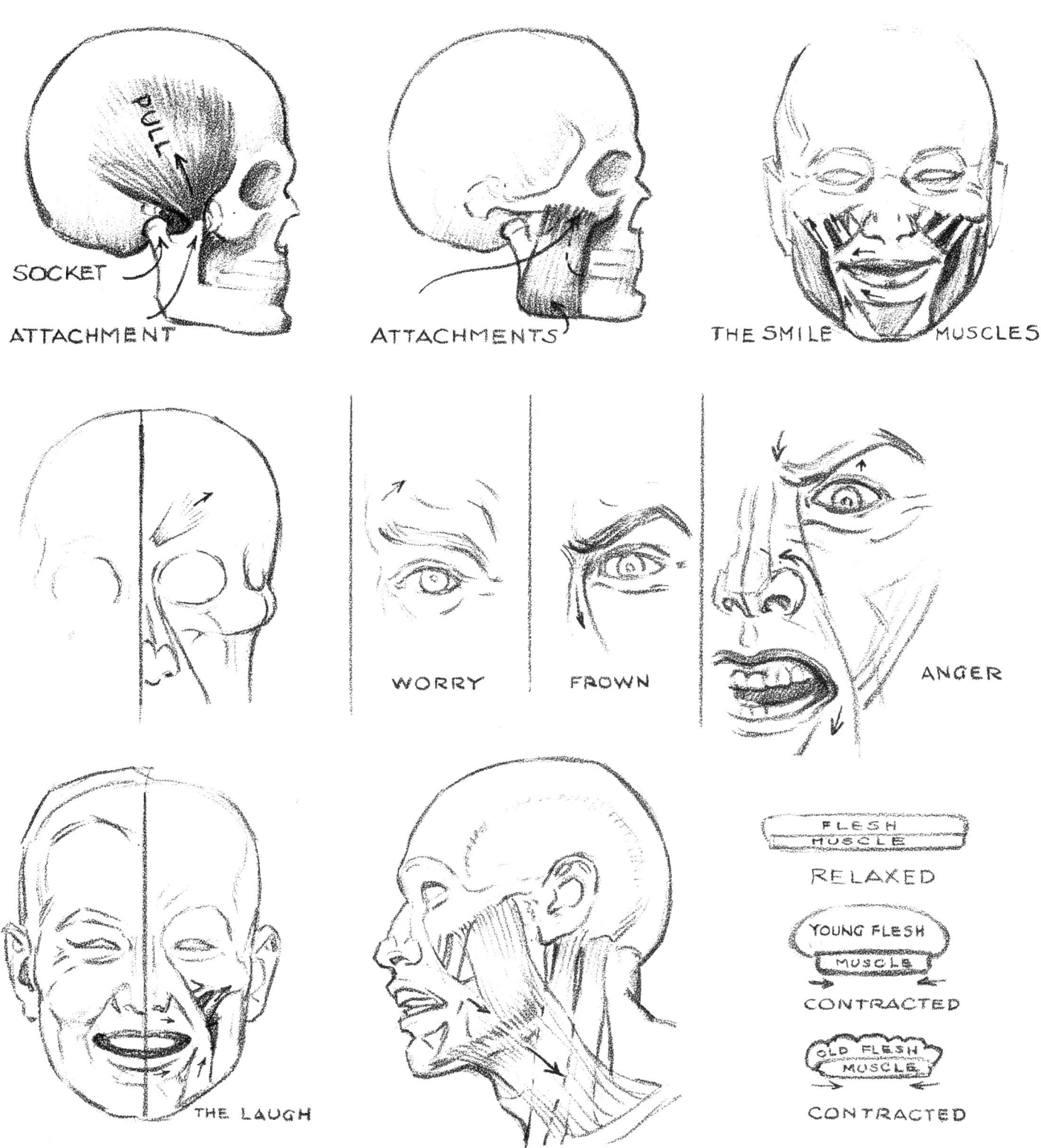

SOCKET

ATTACHMENT

ATTACHMENTS

THE SMILE MUSCLES

WORRY

FROWN

ANGER

THE LAUGH

FLESH
MUSCLE

RELAXED

YOUNG FLESH
MUSCLE

CONTRACTED

OLD FLESH
MUSCLE

CONTRACTED

PLATE 21. How the muscles function

The drawings here, though not very pleasant, are important to the artist if he intends to give his characters expression. The smile is most important in commercial art and advertising. In illustrating fiction you may have to draw an angry face occasionally but the great majority of the faces you will draw are pleasant ones. However, it is much easier to draw a "dead-pan" face than a very happy one. What we want to do is to keep the face that should reflect happiness from appearing as dead-pan or even leering. So study this page well.

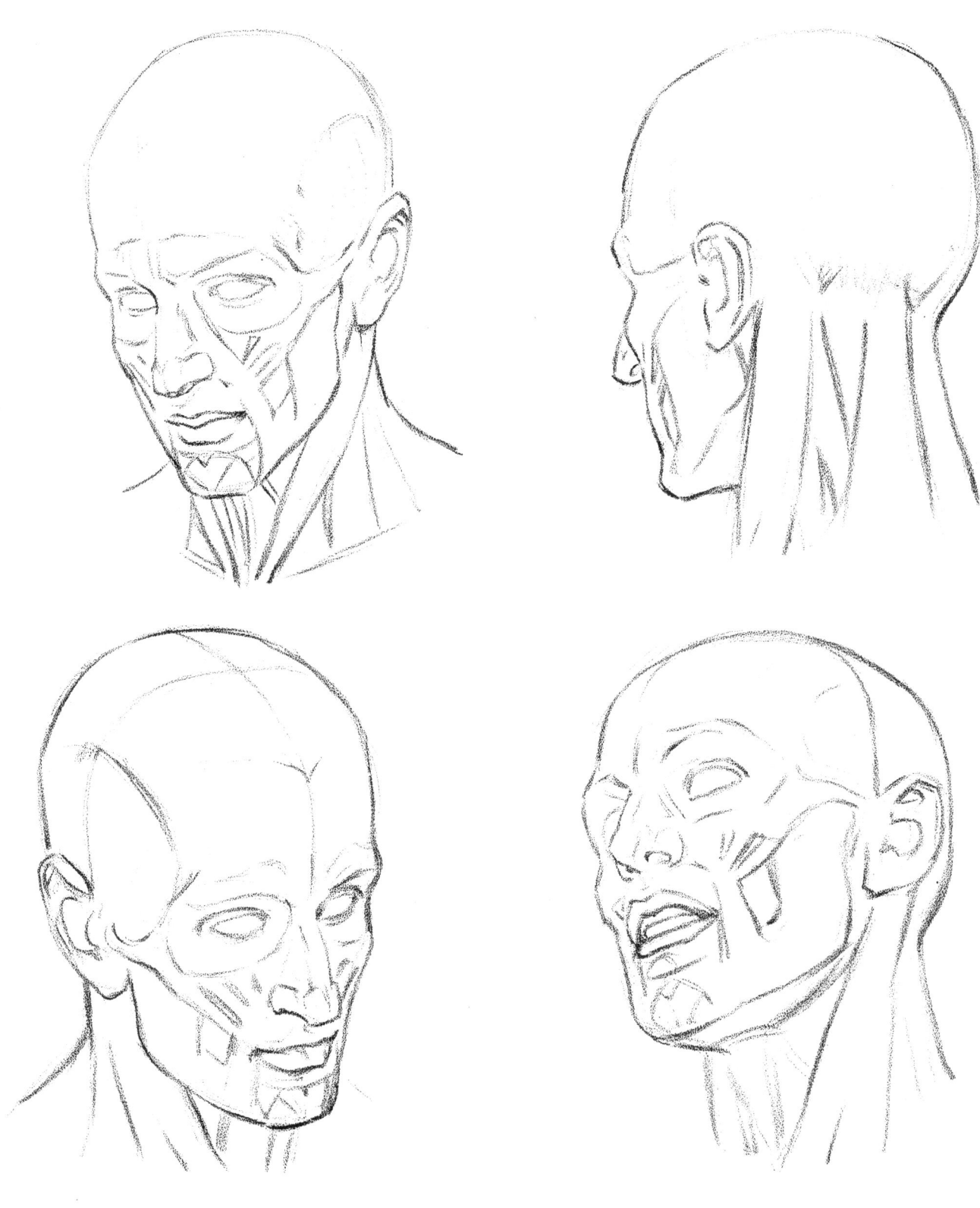

PLATE 22. The muscles from various angles

After you have learned the muscles of the head, try placing them within the head in various poses. Tip and turn the head and line up the muscles to balance on each side of the middle line of the face. You will be surprised to see how easily they will begin to fall into place within the construction plan you have already learned.

48

WHY YOU NEED ANATOMY TO DRAW HEADS

Only a few artists seem to have more than a hazy idea of the anatomy of the head, or of how the muscles function. If faces were expressionless we might manage with only a little of this knowledge. It is argued that we can depend upon photographs for expression. Frankly, many artists do just that. My contention is that one can learn the necessary principles of anatomy in two or three short periods of study, say three evenings. When so little effort is required, why not spend it to learn something that will always be valuable to you.

Every expression is entirely dependent upon a very few muscles lying under and embedded within the flesh. Knowing where the muscles lie and what they do is the difference between guesswork and knowledge. An expression must carry conviction, and it's easier to convince when you know the facts you are dealing with.

For many years I seemed to have great difficulty in drawing smiles. I had taken it for granted that the smile creases began at the nostrils and ran straight to the corners of the lips. Actually the smile creases run well outside of the corners of the mouth and around them and point for a little way toward the side of the chin. This is because the lips lie in an oval-shaped sheet of muscle and the creases form at the outer edges of this muscle. The small ribbon-like muscles which lead down from the cheekbones are attached to this sheet of muscle at the outer edge and cause the smile creases. In some smiles the pull of these little muscles actually causes the corners of the mouth to round out rather than to end in a sharp point. For some reason I had not grasped this in my early studies. The experience proved the value of going back to the source when you are in trouble.

One thing that is important in the smile is the way folds of flesh appear under the eyes. Sometimes these add a good deal of mirth to a smile; sometimes they do not. I cannot tell you why.

Some faces have this characteristic to a pronounced degree, while in other faces it is hardly evident. The difficulty is to make the folds appear natural and a part of the smile rather than to have them look like pouches under the eyes. These folds are easier to paint than to draw, because in painting they may be rendered in light values, but in a drawing we are usually using a black medium, and the folds get too black. The same is true of the wrinkles that show at the outer corners of the eyes in a smile. If these are too black, they look like crow's feet. Many smiles are spoiled because the lines around the nostrils are too heavy and black, suggesting a sneer more than a smile, or making the face look as if it were smelling something unpleasant.

Another valuable hint about the smile is that it shows more of the upper teeth than of the lower ones. That means both a greater number of teeth, and more area of the teeth themselves. The corners of the lips are pulled away from the teeth, causing a hole or dark accent within the corners of the lips. The teeth should never run right into the corners as if they were pressed against the lips all the way around. The pull of the muscles stretches and flattens the lips, but the inward curve of the teeth is still there and becomes even more evident because of the shadows cast inwardly by the lips at the corners. There should be some toning down of the teeth as they go back. The two front upper teeth are the ones to highlight. It is better not to try to model the teeth too much, or to draw lines between them. This again is because almost any line may be too black. The lines between the teeth are really very subtle and delicate. Often the teeth should be suggested rather than drawn in detail—unless you are selling toothpaste. Anders Zorn was a master at painting teeth in a smile.

Plate 23 shows the mechanics of the mouth. At the top are the bones without the flesh. We must always remember that the upper jaw is

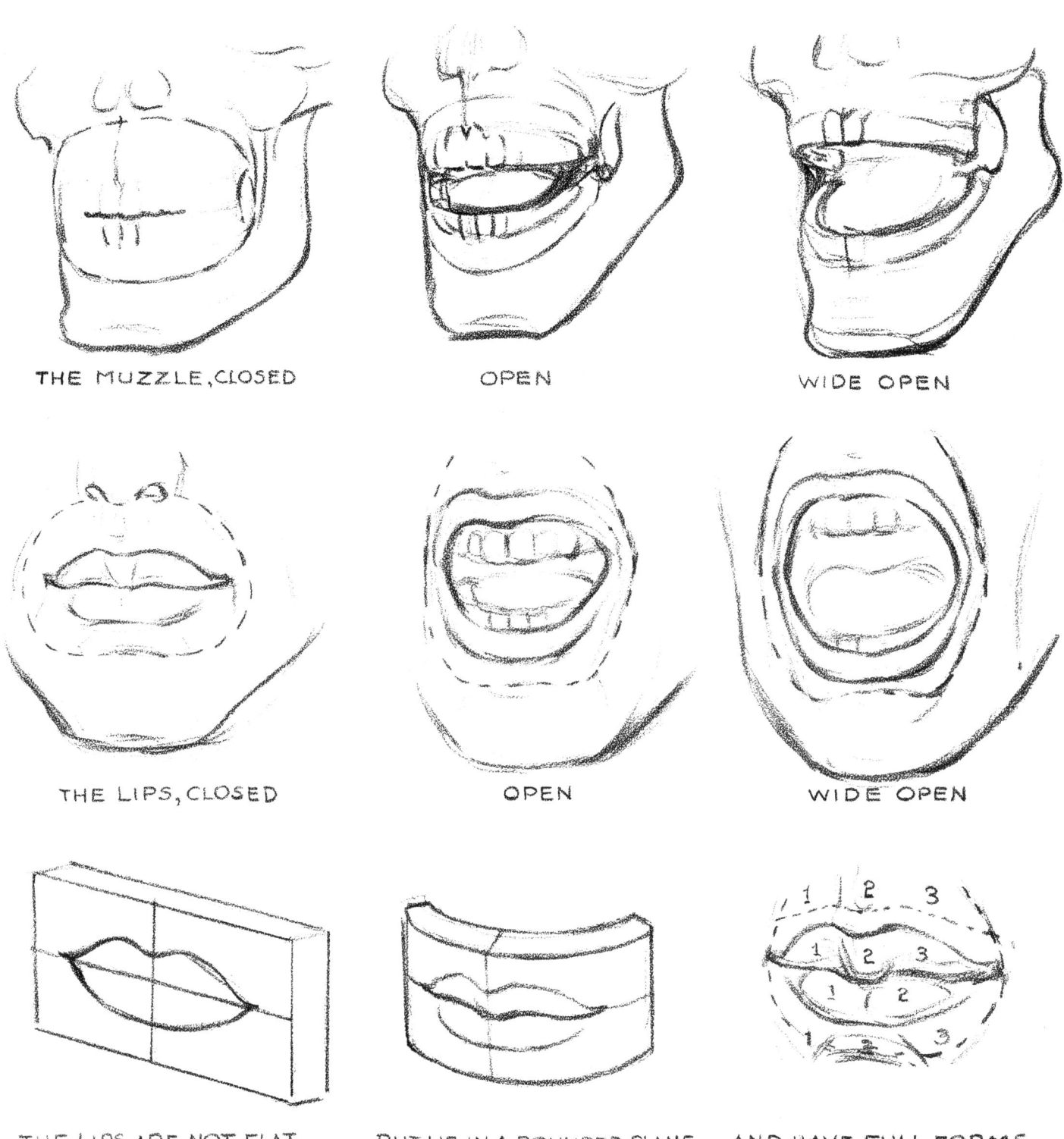

THE MUZZLE, CLOSED   OPEN   WIDE OPEN

THE LIPS, CLOSED   OPEN   WIDE OPEN

THE LIPS ARE NOT FLAT   BUT LIE IN A ROUNDED PLANE   AND HAVE FULL FORMS.

PLATE 23. Mechanics of the mouth

The lips and jaw can hardly be drawn convincingly without an understanding of the muzzle and how it works. Beginners draw the mouth as if it lay on a flat plane. The curve of the teeth in the rounded jaw must be considered, and the fullness of the lips themselves must be felt.

50

fixed in its relationship to the rest of the face, and all the movement takes place in the lower jaw. The curve of the upper teeth remains unchanged and is affected only by the viewpoint. The dropping of the lower jaw may add as much as two inches to the length of the face. When the upper and lower teeth are separated, be sure to compensate by dropping the chin proportionately. And, once again, always consider the roundness of the muzzle all around the lips.

Plate 24 gives you a real look at the eyes. We are too likely to think of the eye as something round (the iris) on something white (the eyeball). Until we analyze the structure we are not conscious of how much the lids are affected by the roundness of the eyeball. The reason is that we see only a little more than a quarter of the eyeball between the lids. But the curve of the eyeball is very evident from corner to corner of the lids. An eye without lids is, of course, a gruesome sight, but we must make these lids seem to lie on the rounded surface. The lids operate almost exactly like the lips. Except in the front view of the face the drawing of one eye is never an exact duplicate of the drawing of the other. When the iris of one eye is at the inner corner, that of the other is at the outer corner. There is a slight bulge of the lens of the eye which travels around under the upper lid. Think of the eyes as two balls working together on a stick. As you turn the stick you also turn the eyes. Think of the lids as the covers over the two balls, in principle like the drawing in the lower right-hand corner of Plate 24. Draw many eyes, first separately, then in pairs. Clip out some pictures of eyes and copy them.

In studying the mouths shown in Plate 25, consider the lips and teeth separately for the time being. Try drawing these mouths, and also get a mirror and draw your own mouth. Move the lips. Tilt your head at various angles. Notice that the teeth are more or less indicated, not by lines between them, but by the gums above and the accents of the dark area below. It is very easy to overemphasize the detail in teeth, so that they do not seem to stay within the mouth. Overemphasized teeth can spoil an otherwise good head.

Noses and ears are shown in Plate 26. Noses and ears are affected by viewpoint and perspective as much as lips are. In other words, these all look the way they do because of the angle from which you see them. You can see why it is so important to establish the viewpoint of the whole head, before we can draw any of these features. When drawing from life it is most important that the pose of the head has not been changed between the drawing of separate features, since that will throw the drawing off completely. A nose must sit within the construction lines of the whole head and over the middle line, or it simply will not look right. The nose and ear should be drawn together, so that their relationship is established. The ear looks very different from the front, side view, or back. See that the nose is at right angles to the line of the eyes and brows. When the brows tip, the nose tips; in fact, everything in the face tips.

Plate 27 gives some examples of laughing and smiling faces. Though these are restricted to line alone, you can feel the muscles operating in the flesh. What I call the sharp-cornered smile is shown on the fellow in the upper right-hand corner. The faces in the middle of the top and bottom rows have a round-cornered laugh. This must come from the subject, for a round corner badly drawn can easily become a leer. Smiles require much study. You can learn a lot with your mirror.

In Plate 28 there are some examples of other expressions, which may give you some idea of how the muscles of the face operate in expressions that are not smiles. The action of the lips can vary a great deal. The basis of most expressions is usually in the mouth. For expressions in cartoons, the cartoonist keeps a mirror handy, since he can assume the expressions he wants more easily than he can explain it to a model.

In using the mirror look for the action of the muscles only; you need not even attempt a likeness of yourself. The mirror gives the artist one big break—he always has a head and hands available to draw from. With two mirrors set properly he can get a side view or a three-quarter view, or make the left hand appear as the right and vice versa.

With expressions, it certainly does no harm to take photographs of a lot of different ones. You can take pictures of your face in the mirror and thus stock up on various expressions for your files. I do not like to see an artist make a crutch of his camera, for I will always maintain that a man can get more into a drawing of his own than any tracing, pantograph, photostat, or projection can give. Photographs have certain distortions that always get into a drawing made from one, unless it is a freehand drawing—and sometimes even then. I think these distortions come from the fact that we see with two eyes, while the camera has only one. The distance of the camera from the subject also has a lot to do with it. Trace a photograph and you will see these things for yourself. Your artistry seems to go out the window, no matter how you try to eliminate that photographic look.

Various types and different expressions are illustrated in Plate 29. I have taken considerable liberty in creating both. It is good training to develop a type, then make several drawings of him showing different expressions. Make him smile, frown, pout, laugh, worry, or whatever else you can. It is really lots of fun, and all the time you are increasing your stock in trade.

In Plate 30 the face has been analyzed to show the structural reasons for the various lines and bumps. When you understand these, you can apply your knowledge in drawing faces of people of different ages, as Plate 31 shows.

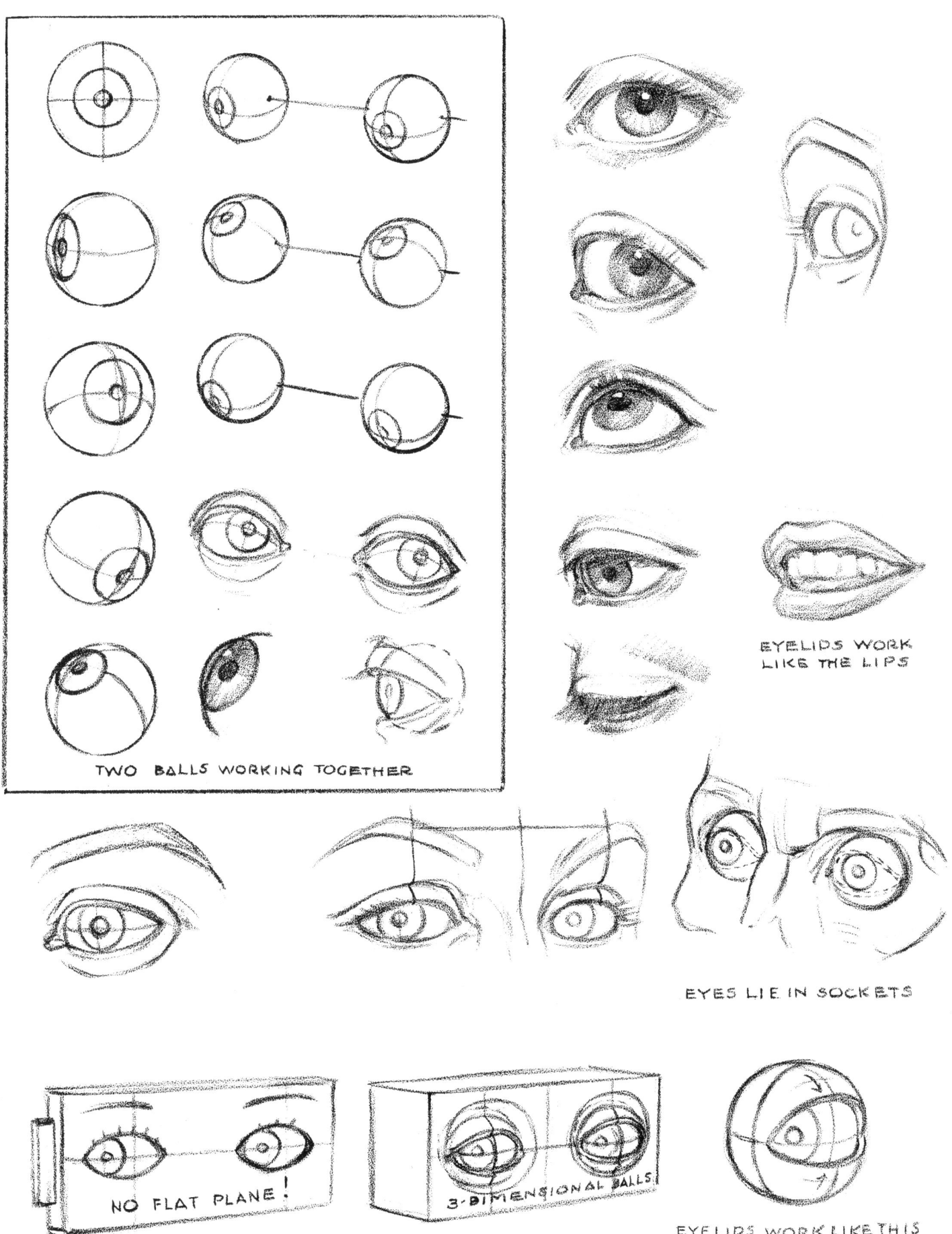

TWO BALLS WORKING TOGETHER

EYELIDS WORK LIKE THE LIPS

EYES LIE IN SOCKETS

NO FLAT PLANE!

3-DIMENSIONAL BALLS

EYELIDS WORK LIKE THIS

PLATE 24. Mechanics of the eyes

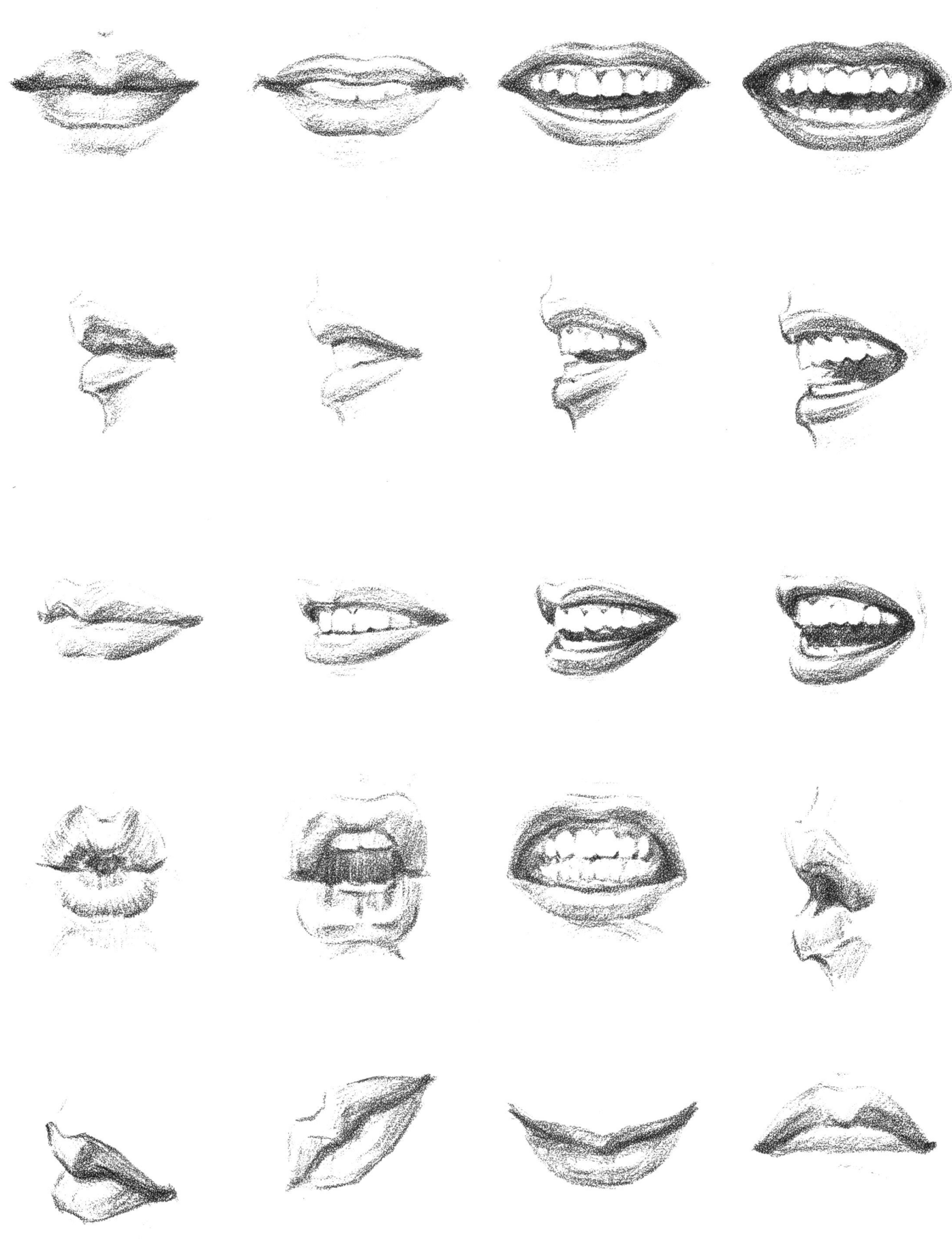

PLATE 25. Movement of the lips

54

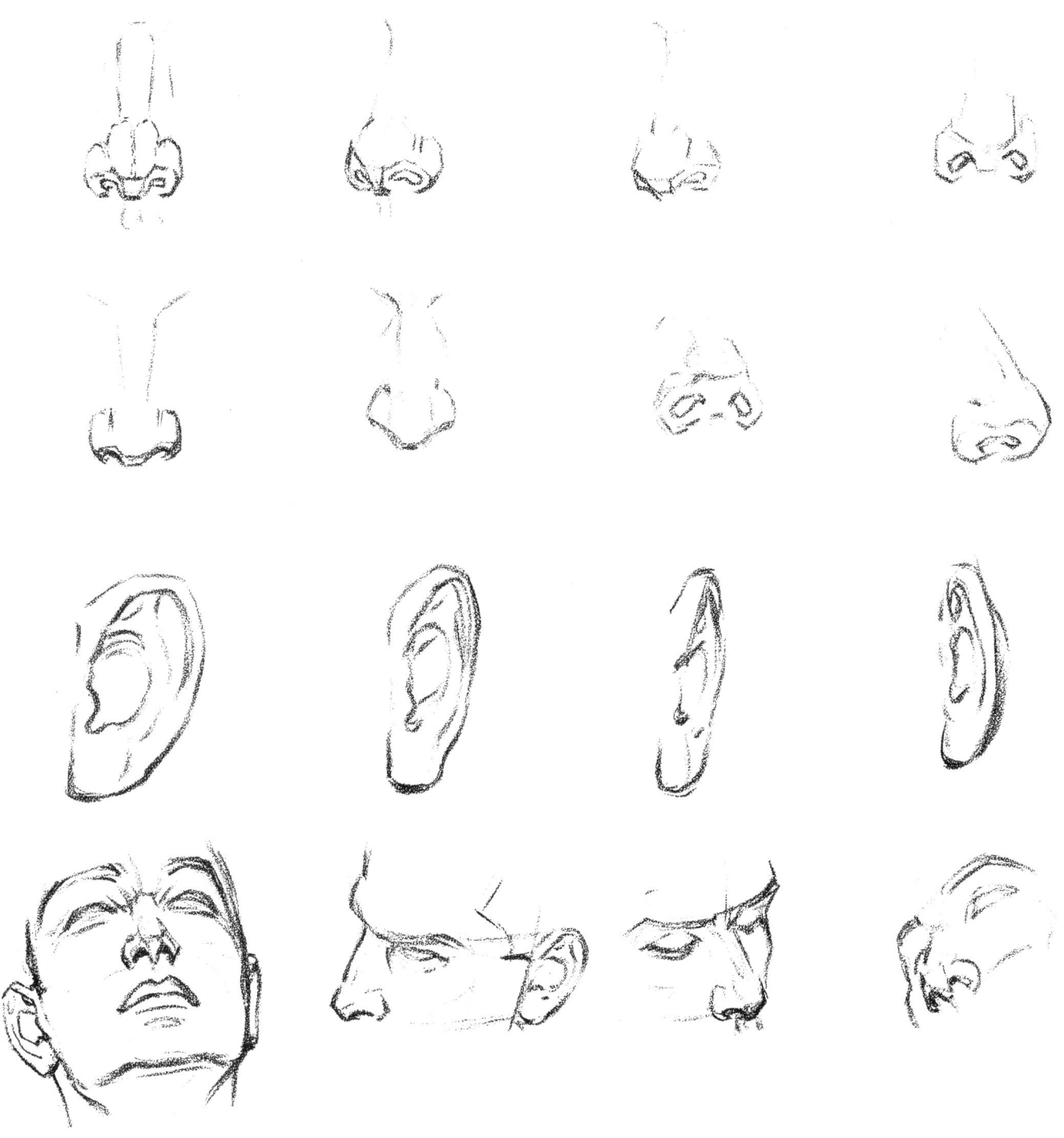

PLATE 26. Construction of the nose and the ears

The appearance of the nose and of the ears is affected by the point of view from which they are drawn. The real problem is much more one of setting them into the construction of the head in their correct positions than one of drawing the actual details themselves. Noses and ears vary widely in shape but not a great deal in basic construction. The nostrils should be set evenly on the line running from the base of the nose to the base of the ear. It is good practice to draw noses and ears from every angle until you are completely familiar with their placement in any pose of the head.

55

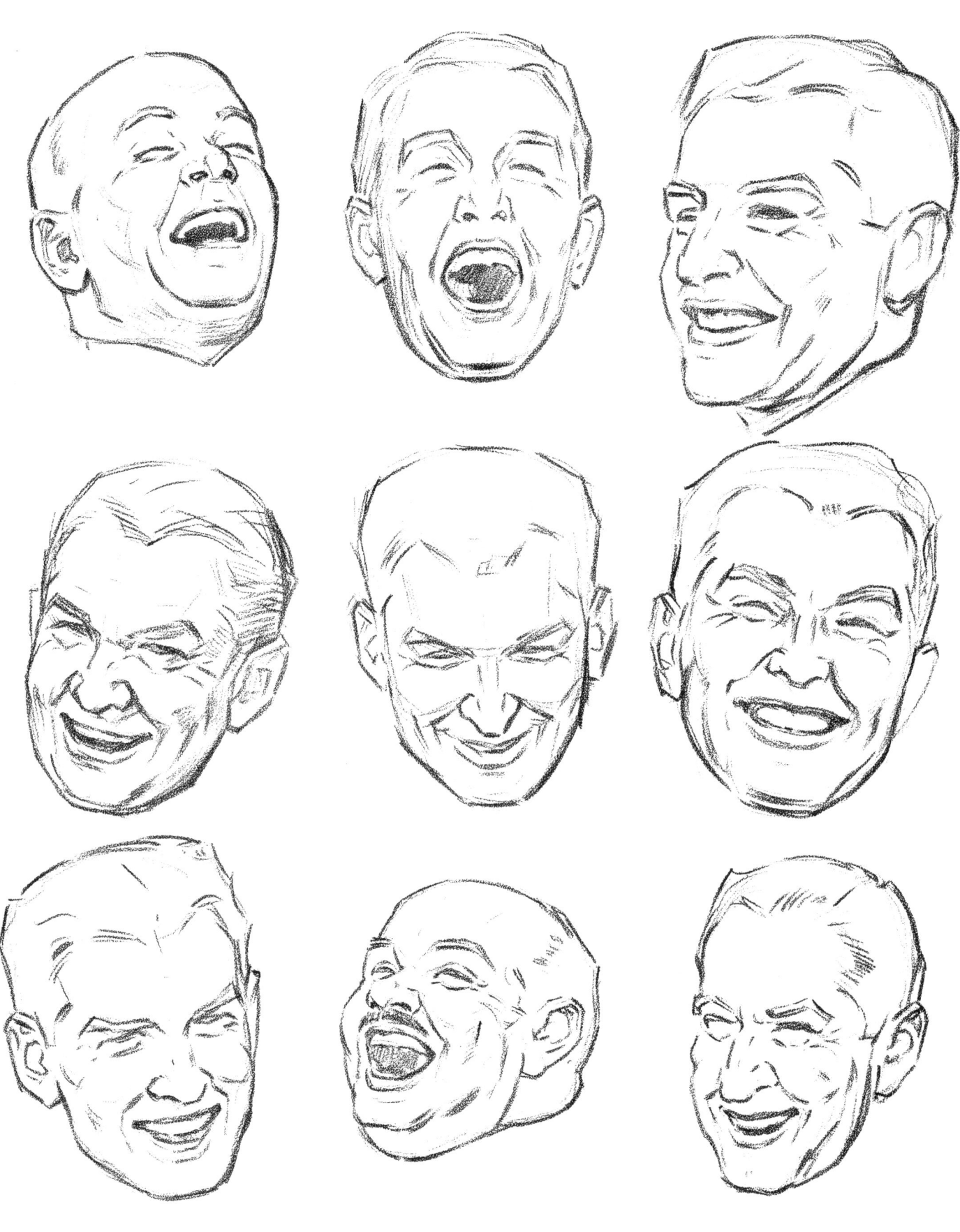

PLATE 27. Expression—the laugh

56

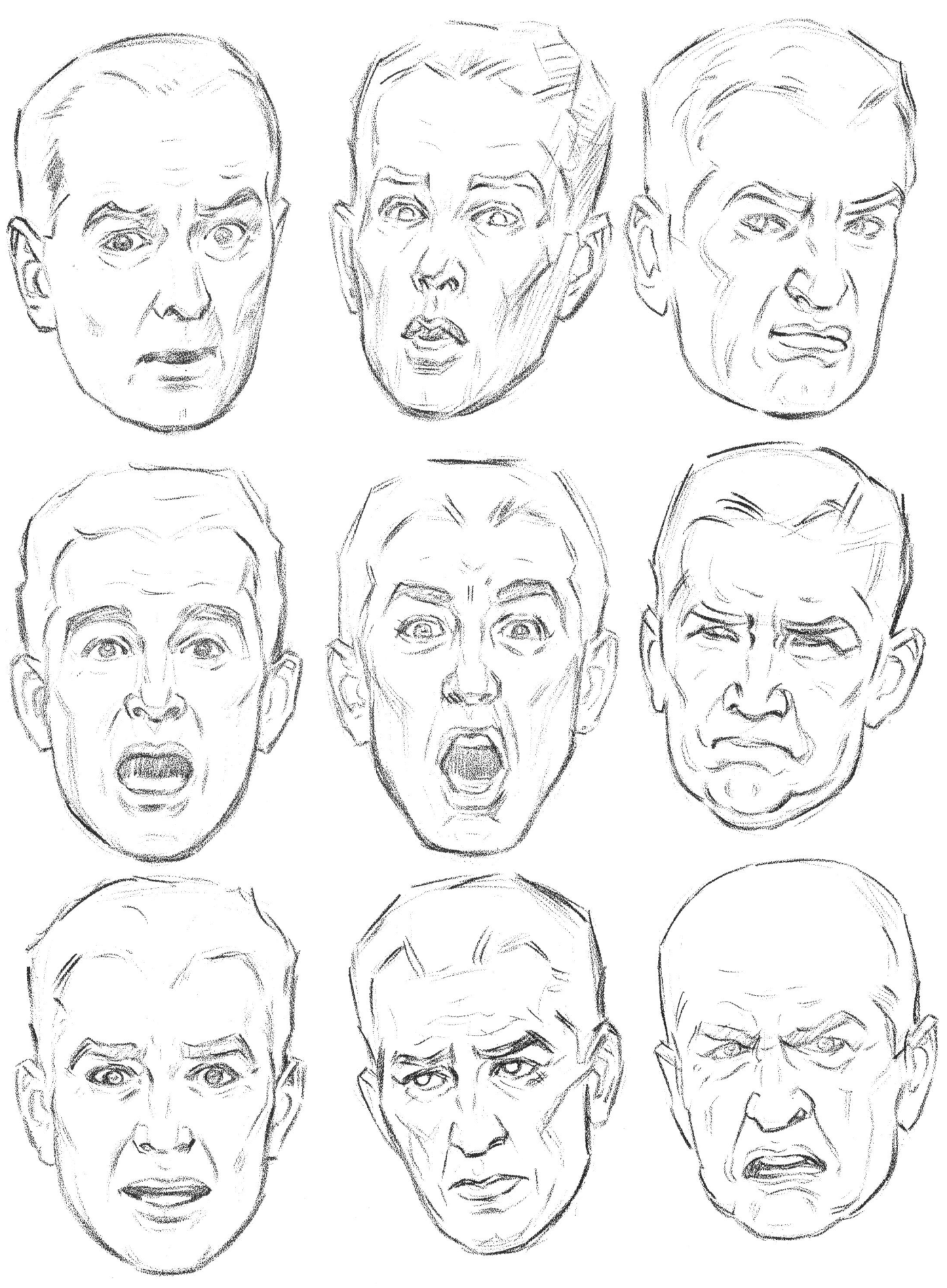

PLATE 28. Various expressions

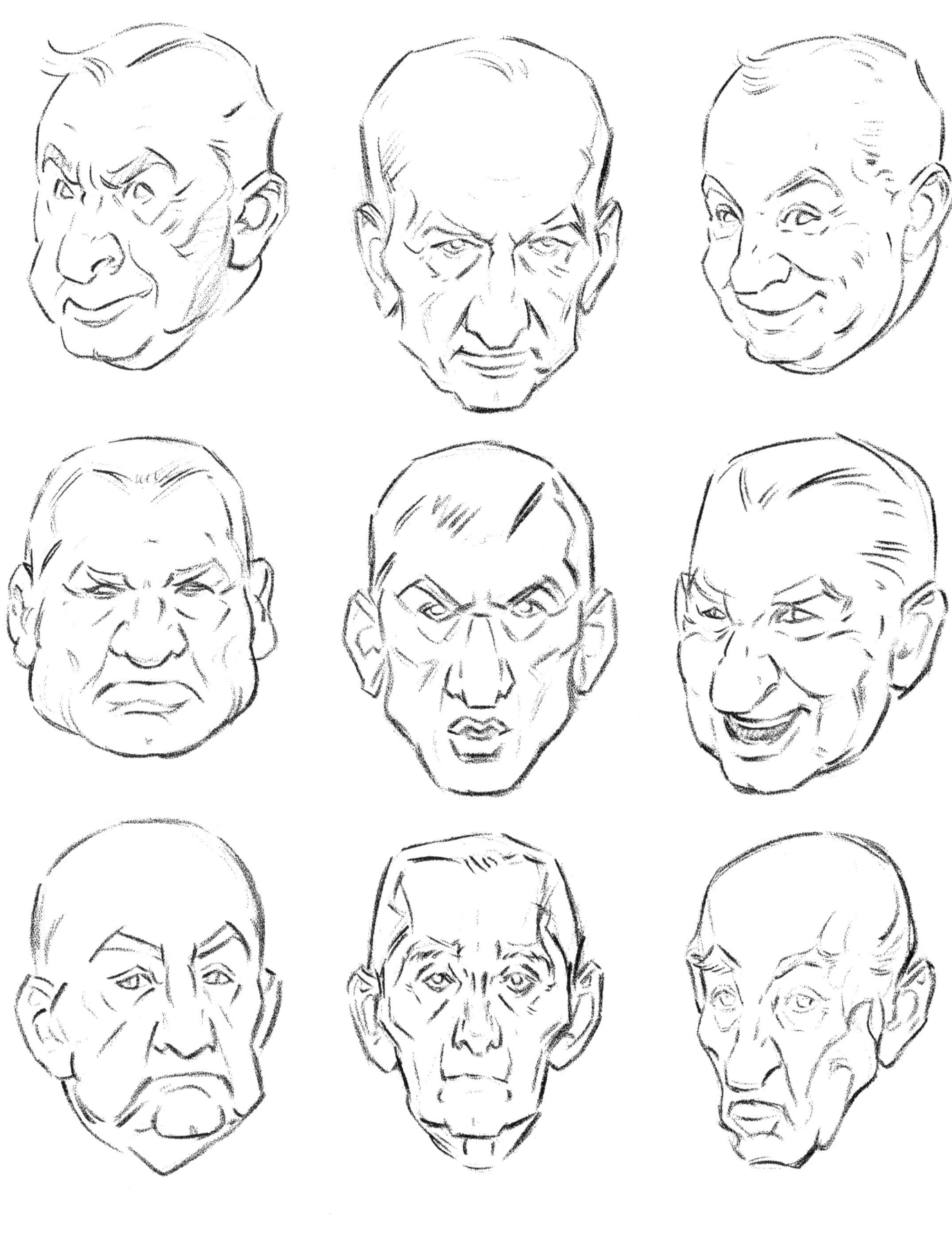

PLATE 29. Characterization through expression

58

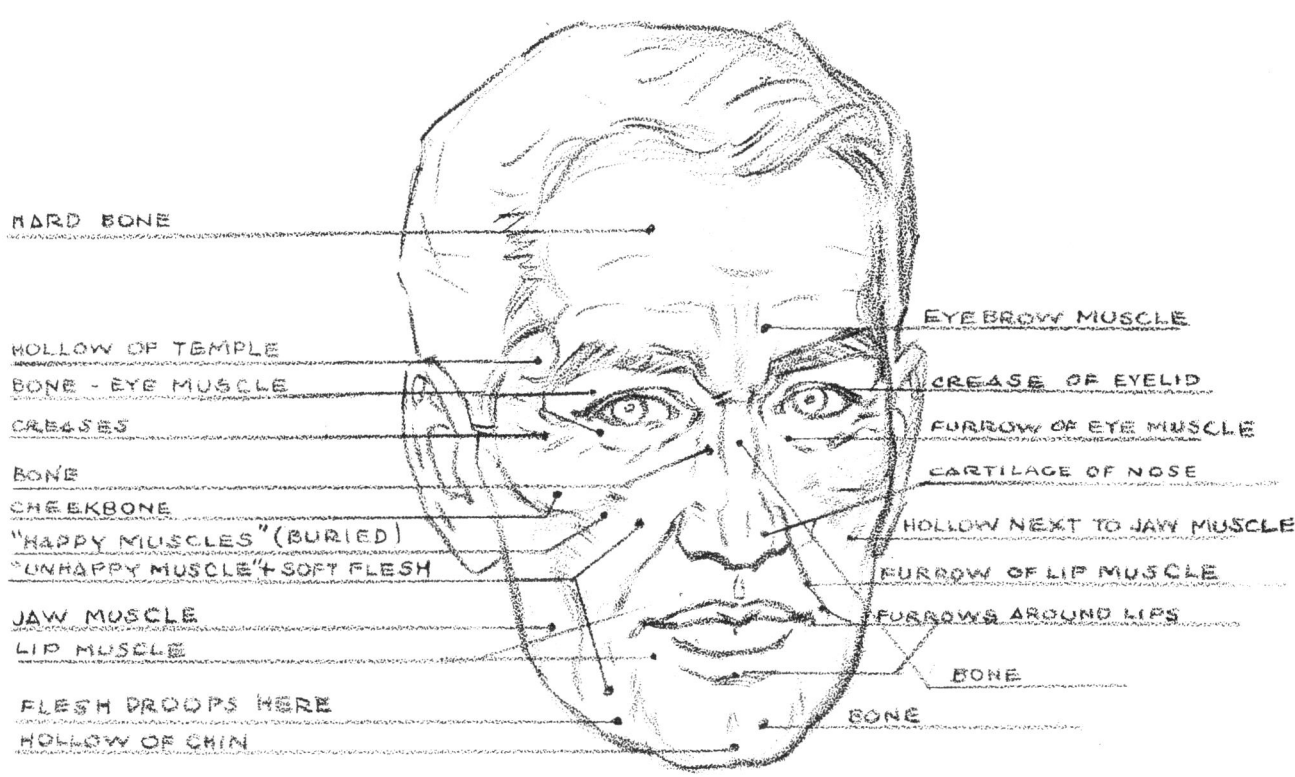

HARD BONE

HOLLOW OF TEMPLE

BONE - EYE MUSCLE

CREASES

BONE

CHEEKBONE

"HAPPY MUSCLES" (BURIED)

"UNHAPPY MUSCLE" + SOFT FLESH

JAW MUSCLE

LIP MUSCLE

FLESH DROOPS HERE

HOLLOW OF CHIN

EYEBROW MUSCLE

CREASE OF EYELID

FURROW OF EYE MUSCLE

CARTILAGE OF NOSE

HOLLOW NEXT TO JAW MUSCLE

FURROW OF LIP MUSCLE

FURROWS AROUND LIPS

BONE

BONE

PLATE 30. Analysis of facial markings

It is not difficult to memorize the size, shape, and placement of the muscles of the face. If you do this, you will thereafter always be able to identify the lines, humps, and bumps in the face. Older people are better than young ones as sources for this information, since the older one gets the more lines and wrinkles develop. We can learn to separate the small wrinkles from the facial lines. The small wrinkles are associated with the shrinkage of the flesh between the muscles, whereas the lines are associated with the edges of the muscles themselves. The small wrinkles of the flesh are seldom drawn or painted since they eventually make a network of wrinkles over the whole face. More important are the forms, and the large creases or lines between them. These are the long creases of the cheeks, those around the mouth, and those over and under the eyes. The muscles are quite pronounced in the male head. When we speak of a strong face, we are speaking mainly of muscle and bone structure.

Only in expressions with raised eyebrows need we worry about wrinkles in the forehead. We can safely leave out most of the wrinkles most of the time and concentrate mainly on the lines, the bones, and the soft forms of the flesh beneath the surface. It is a safe bet that the more wrinkles you eliminate, the better your drawing will be liked. Remember that wrinkles are never black lines on the actual face, but very delicate lines of shadow which can be seen only a few feet away. That is why we can so easily eliminate them and still get a likeness. The deeper creases are evident for some distance, as are the shadows of the planes of the head. Never draw a face as a map or network of wrinkles.

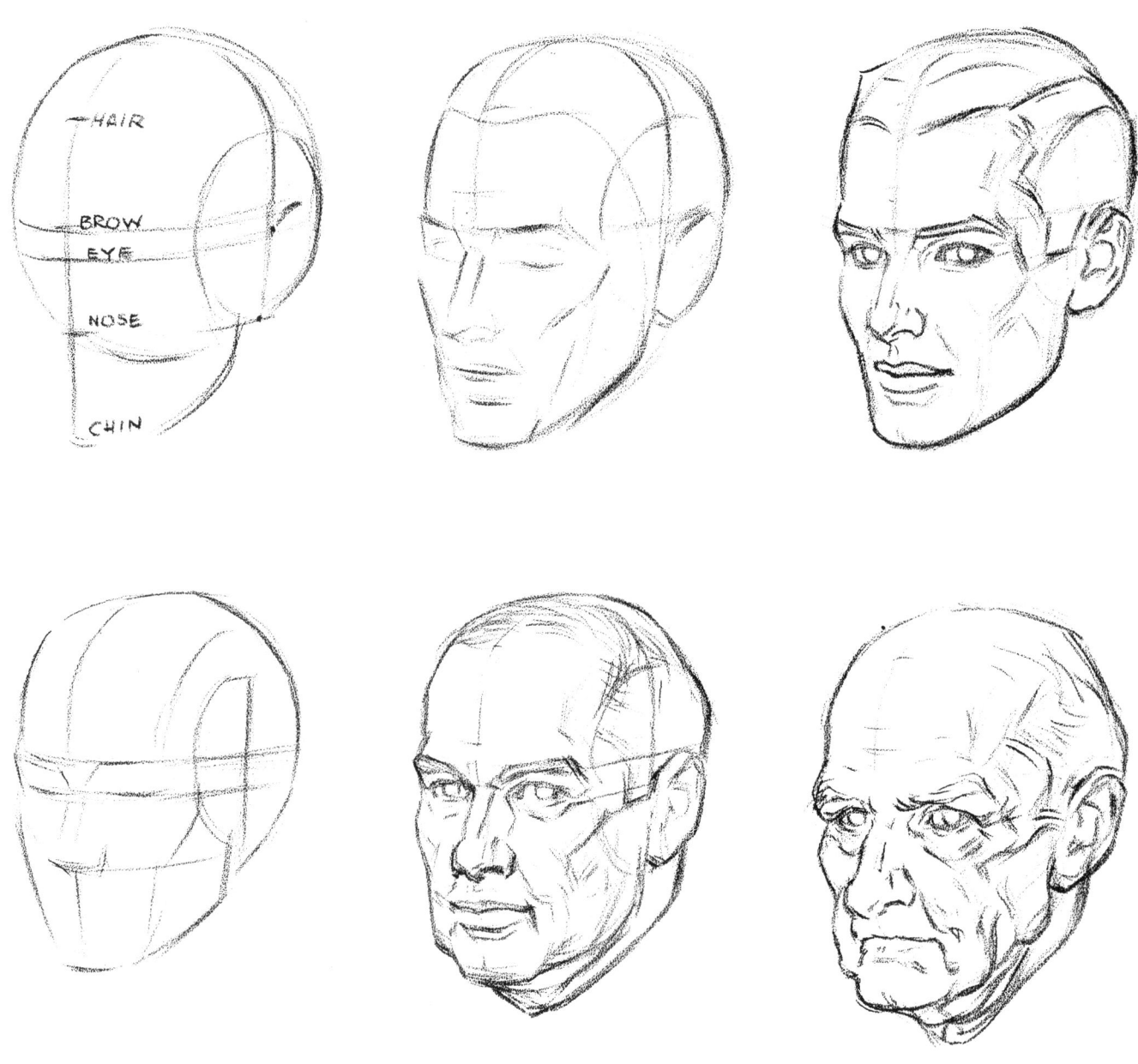

PLATE 31. Drawing faces of different ages

You can easily learn to age a face by adding the forms of the emaciating muscles and the creases that fall between them. The cheekbones, the corners of the jaw, and the bone of the chin become more evident in the aging process. The cartilages of the nose and ears seem to get larger as we get older. The chief change takes place in the cheeks and around the eyes and mouth. The flesh sags at the sides of the chin and along the sides of the jaw. Pouches form under the eyes, and deeper lines at the corners of the eyes. The lips tend to get thinner and move inward, so that more of a straight line between the lips is produced. The lines develop from the corners of the mouth down around the sides of the chin. The flesh above the eyelids droops and the brows seem to drop inward toward the bridge of the nose. A few deeper lines develop across the forehead and between the brows. These can be subordinated, to avoid overemphasizing them. The hair, of course, thins out in varying degrees, so that the hairline moves up and back and there is considerable thinning of the hair at the top of the head. However, we draw the head from the same basic construction.

60

TONE

When we go from line into tone we take a very large step, for tone is the effect of light on form. Although drawing need not carry all the subtlety of tone that painting does, still we must consider values as more or less related. It is better at first to light your subject strongly, or choose a subject that is more or less in simple light and shadow. Shadows are really shapes to draw, shapes that occur over the surface of the form, so that we must consider both, the shape of the form itself and the shape of the shadow on it. Therefore keep the lights and shadows as simple as possible. Hold the light down to one source to begin with. Later on, you may want to introduce some back lighting, but never have both lights shining on the same area. This creates a falsity of lighting, and therefore false-looking form, for form really exists only as light, halftone, and shadow define it. If the light were not there, we would see no form.

In very diffused lighting, we see form much the way we represent it in outline only. If light is coming from all directions the form flattens out, because form turning away from the light source is what makes halftone, shadow, and cast shadow. By cast shadow we mean that the shadow has continued to another plane like the wall, or down across the neck under the chin. Cast shadows have edges of their own, which depend on the direction from which the light is coming. The difference lies in the fact that in ordinary shadow the form has simply turned so far that the light can no longer reach it. On a round form there is halftone before we reach the shadow, and the halftone merges with the shadow. On a square or angular form the shadow sharply follows the edge which cuts off the light, or around which the light cannot reach. The nose casts a shadow in a bright light; the cheeks, being rounder and more gradual as a curve, blend the shadow with the light.

This very blending of light into shadow may make the difference between a good drawing and a bad one. If the edge of the shadow is graduated or blended too much with the light, the drawing loses character; if it is not blended enough the drawing may become hard and brittle. A good way to judge is to ask yourself: Am I holding evidence of the plane or have I lost it? If you have softened the edge so much as to have lost the plane, the drawing is bound to take on a smooth, photographic look. For this reason, planes have to be established when you are drawing from a photograph, since they are not apparent in the photograph itself.

In drawing planes, we can do much to suggest the direction of the plane by the direction of line, without much change in values (see Plate 34). For this reason a drawing can be made to appear very solid, where a wash drawing or painting may lose much of the character. This is a principle which is used effectively in pen drawing, that of making the strokes follow the direction of the plane. It can be used in other mediums that are not areas of flat tone.

I hope the reader will give particular attention to Plate 33, since I consider this page one of the most important in the book. The drawings here encompass practically all the material offered up so far in this book. Here we have the plan of construction, the anatomy, the planes, and the finished rendering combined in a single pose of an individual head.

In addition to studying this page carefully, find some material of your own. See if you can render in separate drawings what you believe must be the correct proportions, anatomy, and planes of the particular head. You will learn more by doing this than by copying a hundred heads as they appear in your copy material. It will definitely point up anything lacking in your knowledge thus far. When you have, to your satisfaction, worked out the several stages, paste them on a sheet and hang them up in the place where you work, as a constant reminder. If you have worked them out convincingly you can well take pride in the fact. They will be of interest to anyone, for through them you have stated your knowledge in no uncertain manner. They serve to help you memorize the qualities which should go into a well-drawn head, but which, of course, could not be incorporated into a single drawing with each stage in evidence. In the finished drawing, I believe you will feel this background of effort, which I hope will convince you that drawing heads is more than mere copying.

Plates 35 through 39 may help you in the matter of technical rendering, though it is my feeling that technique should be left very much to the student himself. The problems of proportion, anatomy, and planes are basically the same for all of us, but technical solutions of those problems are, to a large extent, an individual matter.

Unfortunately, the student is usually unable to see many good examples of head drawings, because so few are published. In the past decade there have been few men in the field good enough to have their drawings published regularly, aside from the fact that many artists' ability to draw the head is concealed by their use of mediums. I would like to call attention to the work of William Oberhardt, who stands almost alone in drawing the head. I hope the reader may at some time come across a few of the many drawings of his that have appeared in publications. The schools in England seem to have produced many more fine examples of head-drawing than those in America have. I think this is because the young American artist tends to turn to photographs for material before he has any real knowledge of the head. The drawings in this book are offered humbly, since there are many draftsmen whose skill exceeds mine, but because of the lack of helpful books on the subject, I submit whatever I have to offer hopefully.

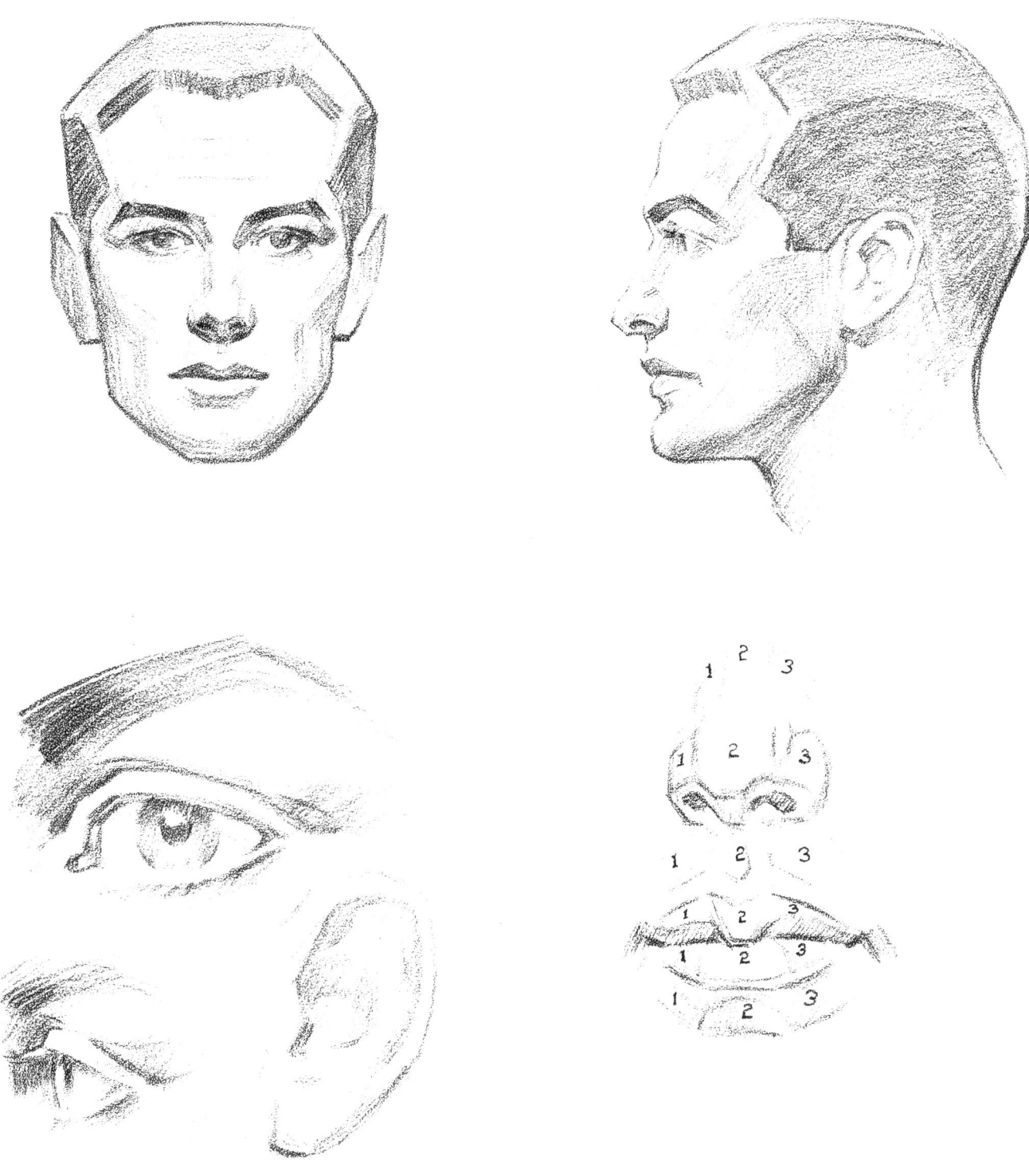

PLATE 32. Modeling the planes

As a basis for learning to show light on form, turn to Plate 9 and make a drawing of the planes of the head as shown there. It will help you a great deal with the material to follow. Let us understand that we can depict solid form only as it appears in light, halftone, and shadow. The shadows get darker as the form turns away from the light. A single light is always simple to draw, for more than one light cuts up the shadow tones, making everything more complicated. Think now in terms of flat areas in varying tones, and forget surface wrinkles entirely.

63

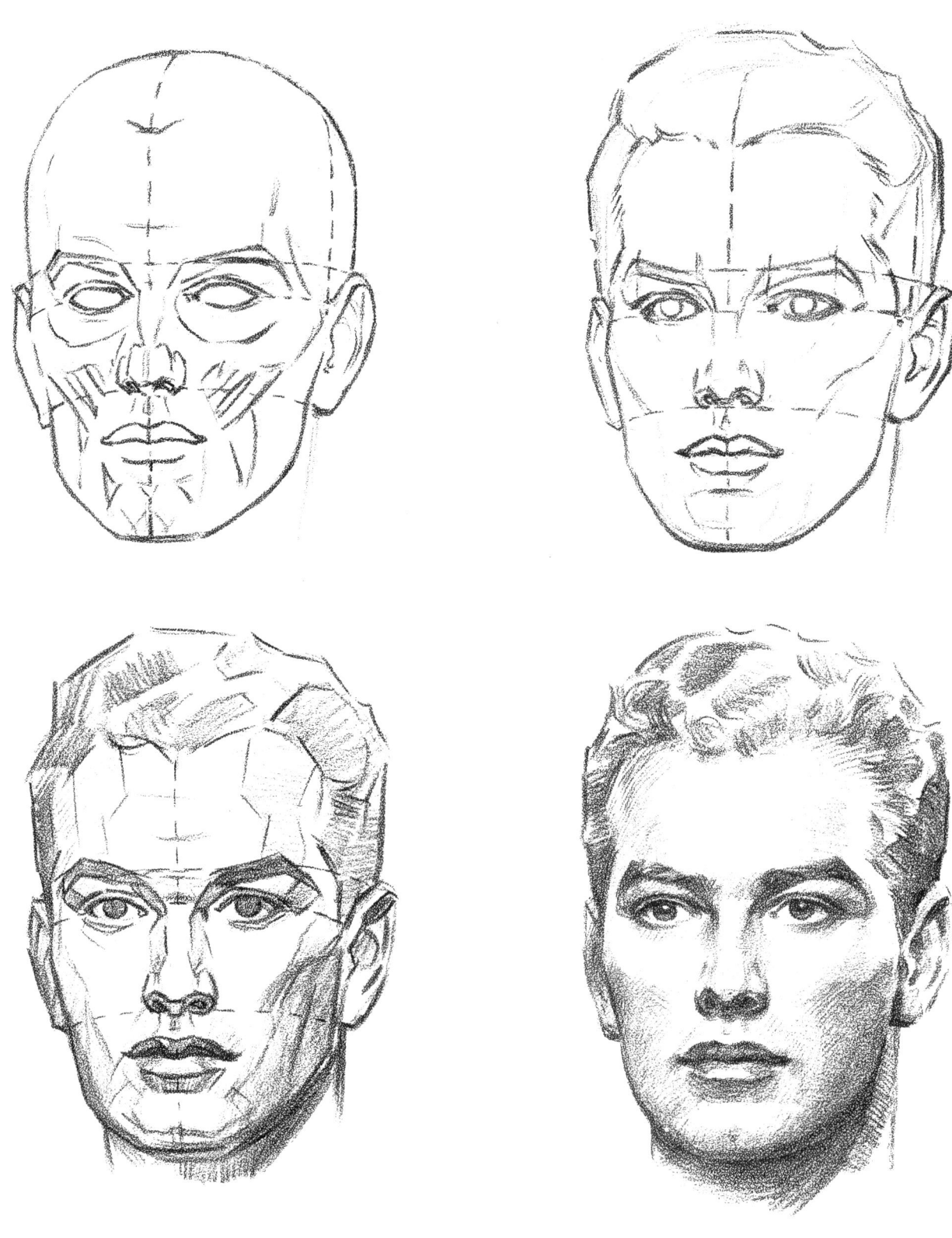

PLATE 33. Combining anatomy, construction, and planes

This page is one of the most important in the book, since it shows the stages of drawing a head from the anatomy and construction, through the outline, to the planes and the final completion of the drawing. It would be impossible to follow without considerable study of the preceding information, not in order to copy this head, but to draw one yourself. Study this page carefully; you will find it invaluable for reference.

64

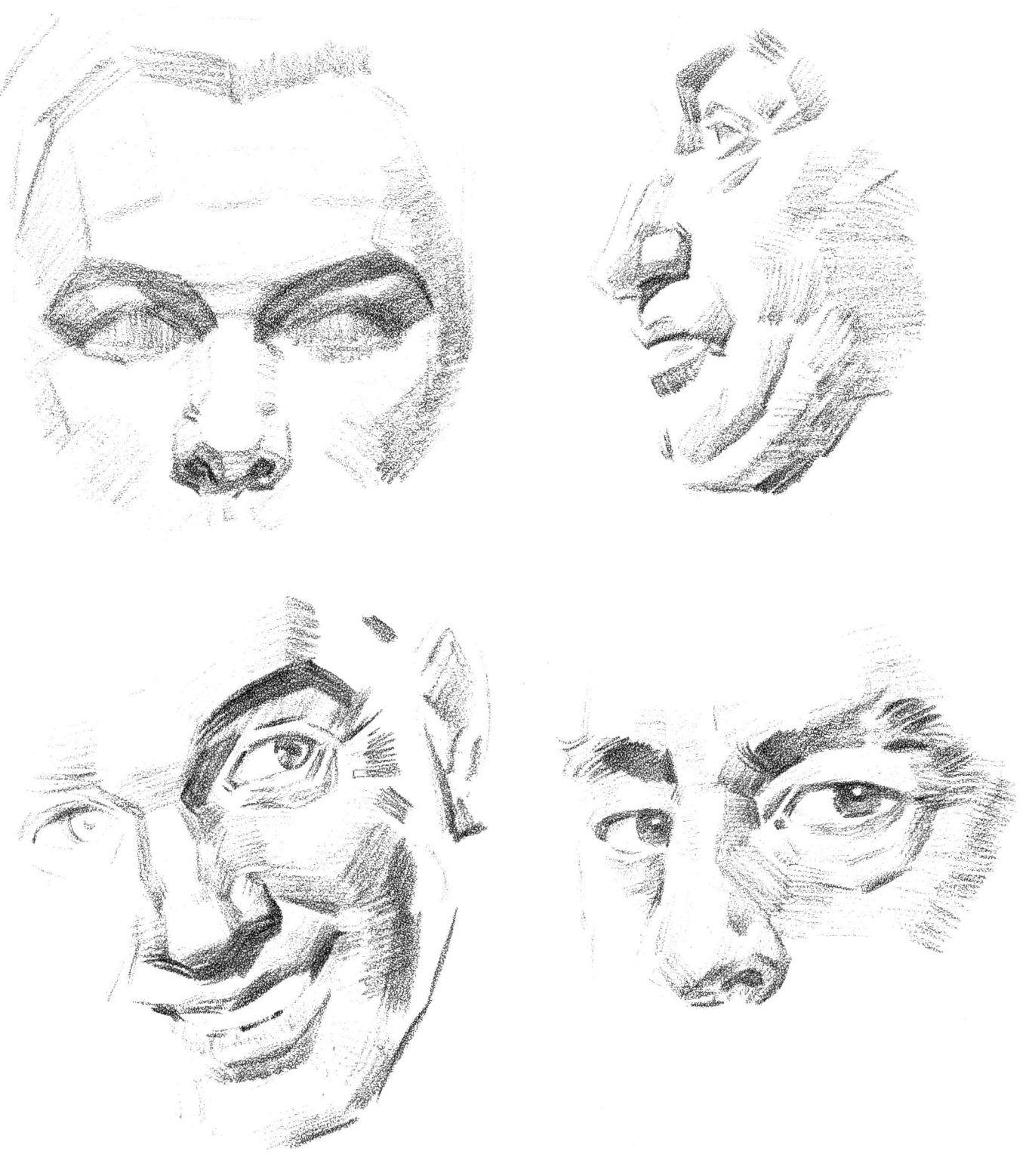

PLATE 34. Building tone with planes

This page shows how the planes may be treated as straight flat surfaces, each carrying its own value between light and dark. The very light planes should have very little tone and be treated very delicately. By directing the stroke, you can make the plane turn without changing the value more than slightly. You get more solidity if you make all the planes in the light a little lighter than they appear, and those in the shadow a little darker.

65

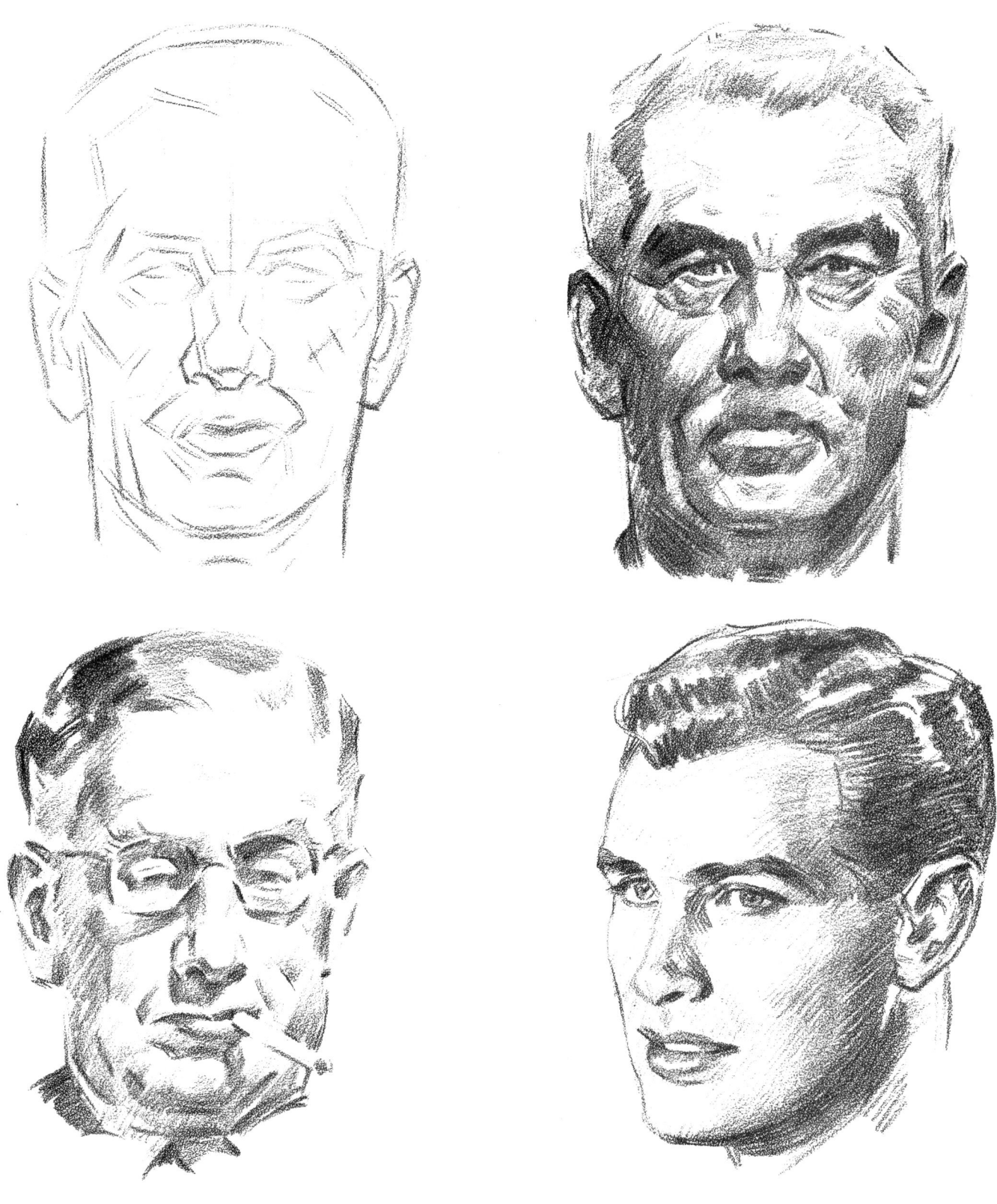

PLATE 35. Every head is a separate problem

Every head is an individual assemblage of shapes, lines, and spaces. Because of the variations of skulls and features, together with variations of spacing, millions of combinations occur. Forget every other face and concentrate on the one you are drawing. Accent the individual forms wherever you can. Start drawing real people, and collect clippings and photographs to practice from. Don't be tempted to trace; just draw.

66

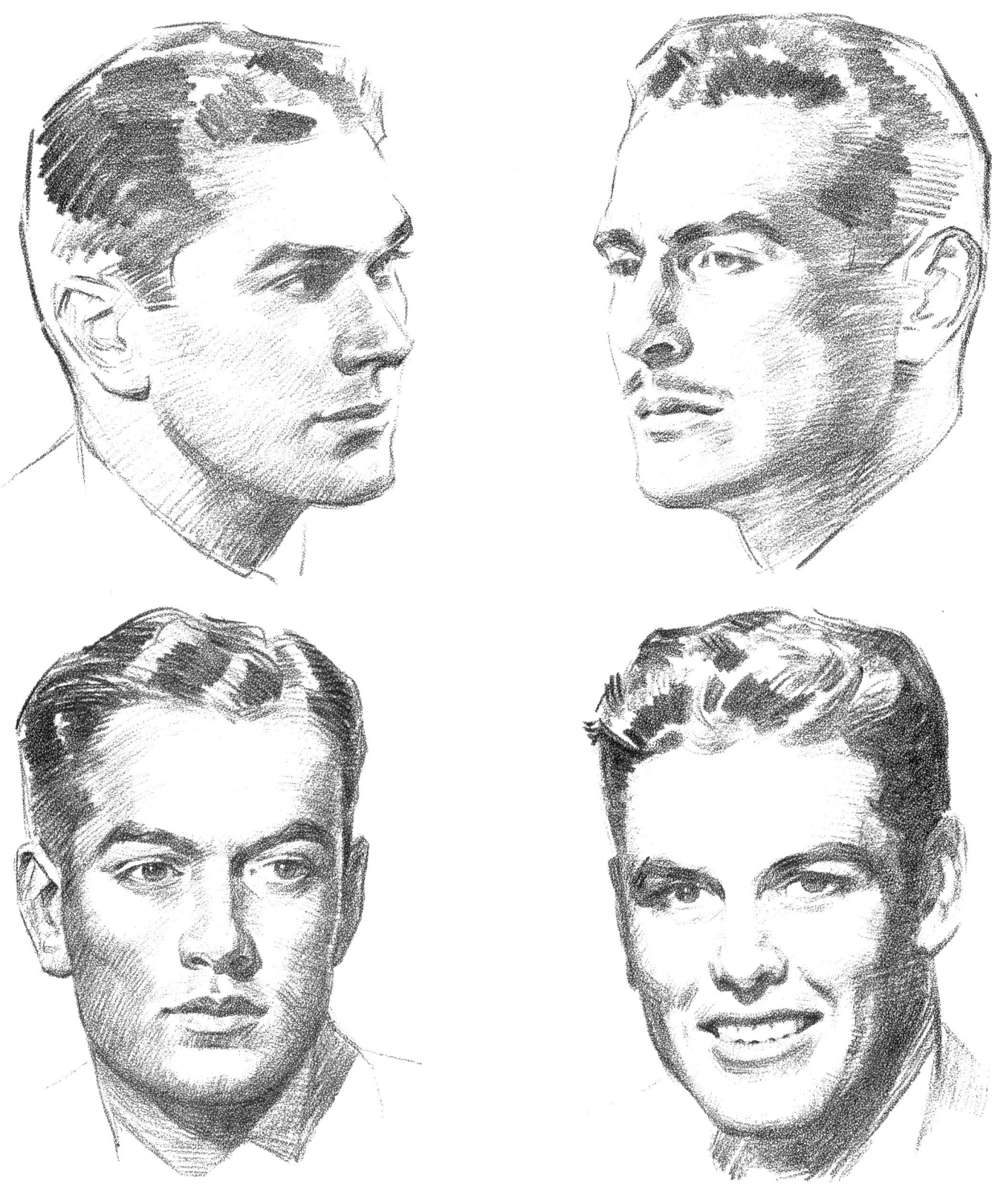

PLATE 36. Types of character

The character in a head is the result of the individual bones and muscles, as they are shown by careful construction and spacing. But the beauty of a drawing will always be in the way you use line and tone and the interpretation of light and shadow on the forms. You may experiment in your own way and develop your own approach and technique. Sometimes an unfinished study is more attractive than the completely executed drawing.

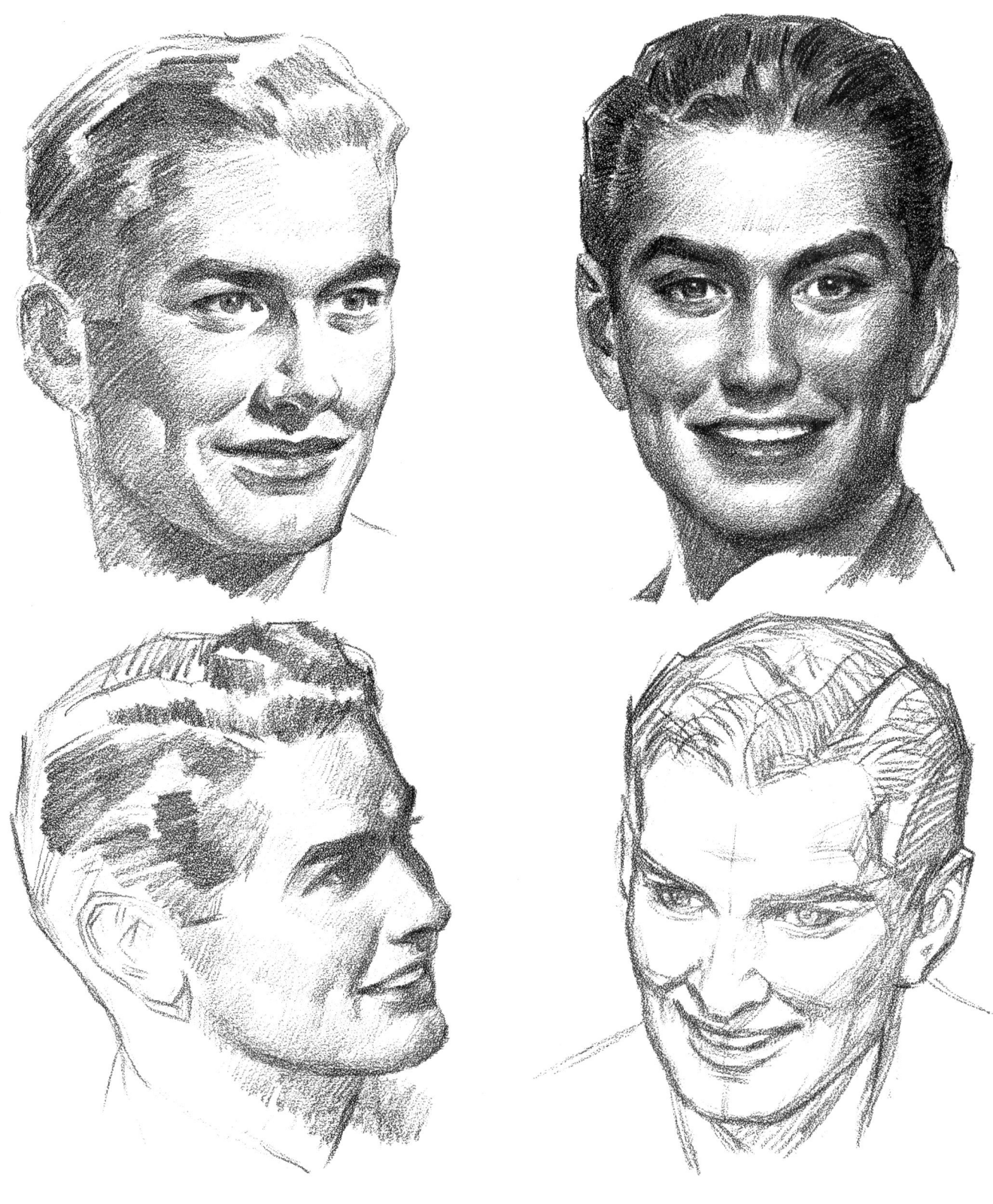

PLATE 37. Smiling men

Smiles that radiate happiness are difficult for any artist. They are much easier to render in an outline drawing than a tonal drawing. If your drawing of heads must provide an income you will do well to practice drawing smiles from clippings, since a model can rarely hold a genuine smile for very long. Study particularly the forms around the corners of the mouth, and the forms of the cheeks.

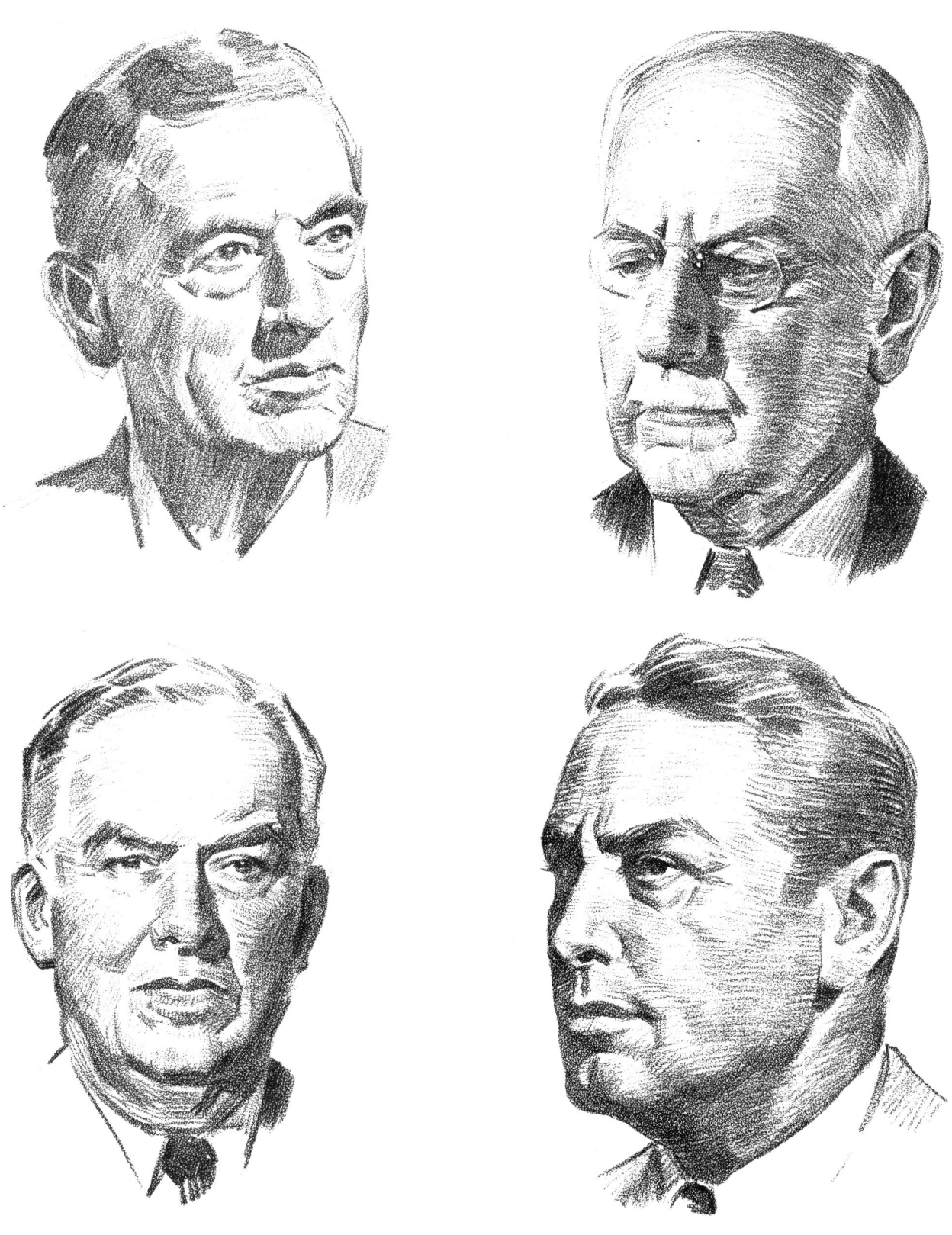

PLATE 38. Older men

The faces of older men give the artist more to "get hold of" in the way of forms and lines. Note, however, that in the faces on this page most of the surface wrinkles have been eliminated and only the main lines and forms stated. The impression of age is maintained without the incidental and insignificant wrinkles.

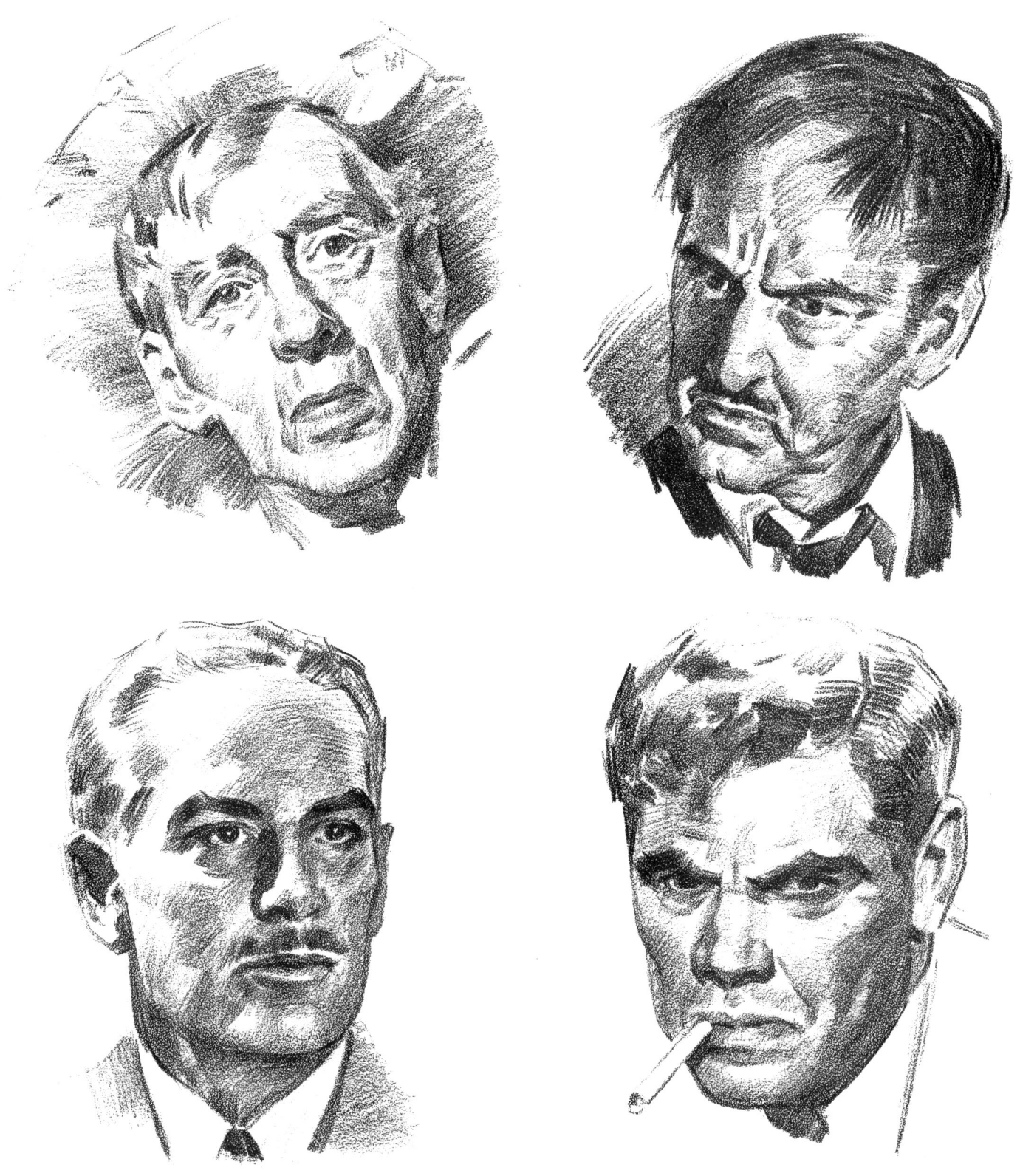

PLATE 39. Characterization

Here construction, lighting, and expression are combined. This is characterization, the way a face looks at a given moment. Expression is really no more than a distortion of the relaxed forms of the face. Such distortion causes movement in the muscles below and consequent change on the surface. Therefore it is important to know how those muscles move (see Plate 21).

70

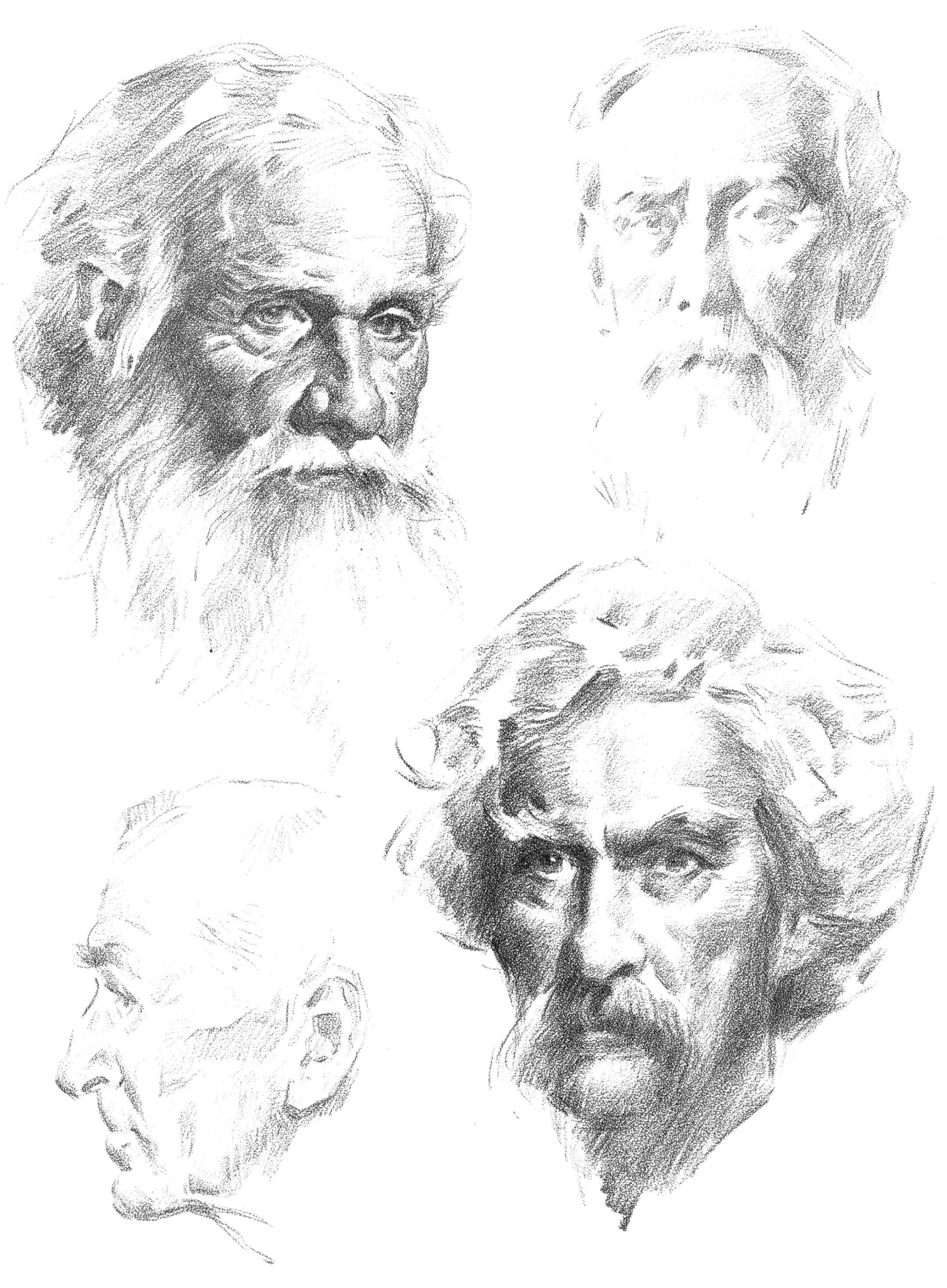

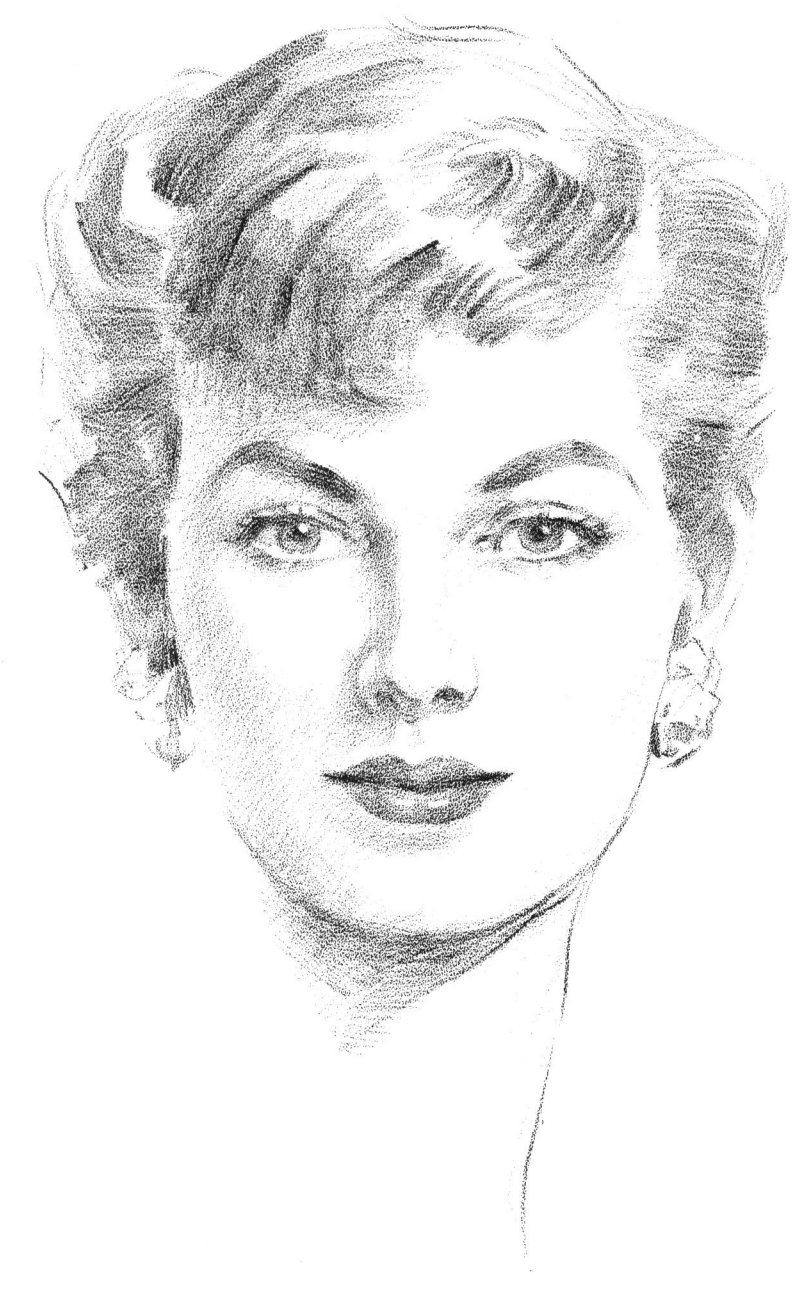

Part Two: Women's Heads

Part Two: Women's Heads

IN AMERICAN ADVERTISING and magazine illustration the ability to draw women's heads effectively is the greatest boon to the pocketbook. While commercial art has many departments, no other is quite so lucrative. This skill opens the door of advertising agencies, editorial offices, and calendar producers as nothing else can. Portrait drawings are much easier to sell than finished paintings, since the price is much lower. Drawings, nicely framed, can be hung anywhere in the house, while painted portraits are more or less restricted to the space over the living-room mantelpiece. A man often prefers a nicely done drawing of himself or his wife or children to an elaborate painting. Fortunately, the artist can make such drawings inexpensively, in much less time than a painting takes, and he can well afford to keep his price within the normal family budget. There are possibilities in portrait drawing which should not be overlooked. It is pleasant work. It can be part-time work, and it is remunerative. If you do studies for one family, others become interested. Such studies make attractive pictures for dens, halls, offices, and other places where furnishings are not elaborate. There is hardly a mother who would not like to have sketches of her children. There are many artists in this country already doing very well at making portrait drawings. The prices usually range from $50 to $150 and even higher, which is not too bad for a few hours' work. These sketches may even be done from camera studies with the personal ability and knowledge added to the photographic appearance.

When you are drawing women's heads, be sure to use freedom and looseness of technique in representing the hair. Usually simple planes are much more effective than the photographic representation of every strand or curl. Another important quality, which I have pointed out earlier, is a blocky effect. The camera sees everything in its roundness; the artist sees its rhythms and its angles.

For some reason a little masculinity is much more tolerable in a woman's head than roundness and femininity is in a man's. The fashion experts seem to pick the lean-faced, angular-jawed, and bony types of models oftener than the purely feminine types. It may be that to get the rest of the figure slim enough to go on a fashion page, a bony face is required. Somehow the appearance of bone in the face does seem to give more character to a woman, just as it does to a man. Perhaps most of us admire leanness more than plumpness because leanness is hard to attain and keep. At least in that we have changed since the days of the old masters.

All this means that in drawing women we still must be conscious of planes, even if we do not stress them as much as we do in drawing men. Plate 42 shows a man's head contrasted with a woman's head in the same pose. Note that the feeling of planes is evident in both, but more stressed in the man's head. Note also that the handling of the mouth and nose is more delicate in the drawing of the woman than in that of the man. If I do nothing else here I want to impress on you that smoothness and roundness are basically associated with the female, and squareness or angularity with the male. The degree to which you emphasize the one or the other in either case is determined by personal feeling about your subject. Plate 44 demonstrates how blockiness may be applied to women's heads.

Plates 45 and 46 are technical examples of women's heads which you may find of some interest. Plates 47 and 48 are sketches in which both roundness and squareness have been felt. I suggest that you make a great many sketches of this kind from life and from the wealth of material provided in magazines.

75

Plates 49 and 50 deal with the characteristics of aging. Drawings of elderly women are the one place where fat seems permissible. Everyone loves a plump grandma.

It is in drawing older women that your knowledge of anatomy is most evident. Younger women strive to keep the anatomy of the face pretty well covered up, and we please them most by doing the same in drawings. But sooner or later wrinkles and creases will come. We can subordinate the wrinkles, but we must take the forms very much into consideration. New forms have developed in the cheeks; indications of the way the muscles are attached in and under the flesh have begun to show through. Bone comes to the surface, for it is no longer so firmly covered by flesh. Pockets form between the muscles for the same reason. Soft flesh stands out in little lumps and begins to drape somewhat toward the chin. We can be kind about it and not put too much emphasis on the aging process, but to ignore it entirely would be to lose both character and likeness. There is beauty in maturity and even in old age. By then character shines through, and there is no graciousness and charm greater than that of an elderly woman of character, who has put away most of the foibles and frivolities of youth. Be kind in your drawings, but do not fabricate. Insincere work does personal harm to your reputation, and that is more important to you than any single drawing of any face in the world. Study the aging process, be thoroughly familiar with what happens, and then treat it tenderly.

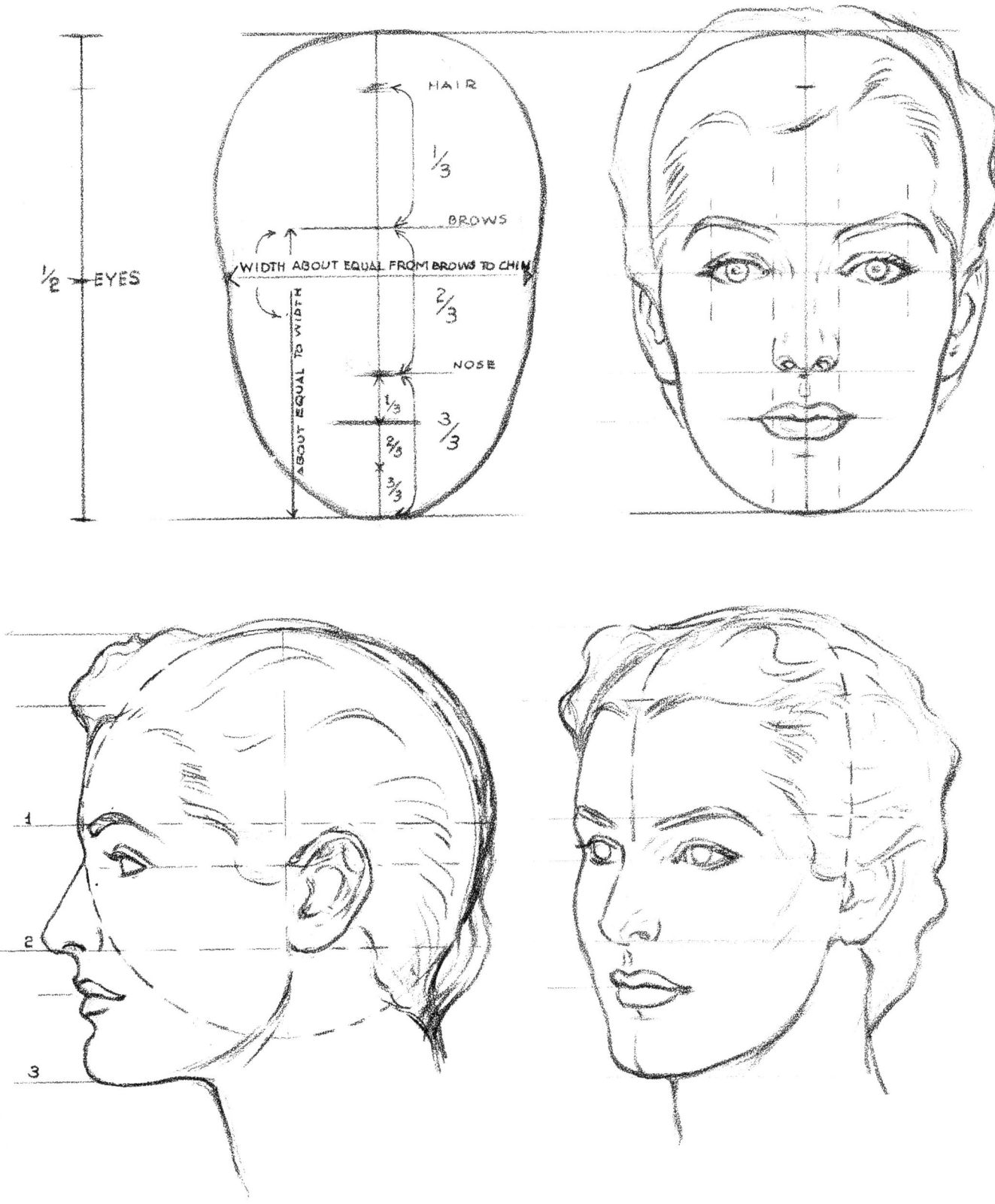

PLATE 40. Constructing the female head

The over-all proportions of the female head vary only slightly from those of the male head, but the bone and muscle structure is lighter and less prominent. In commercial art feminine types with rather firm jaws seem to have more appeal than do the very rounded. Women's eyebrows are usually a little higher above the eyes than men's are. The mouth is smaller; the lips are more full and rounded, and the eyes slightly larger. Do not stress the jaw and cheek muscles.

77

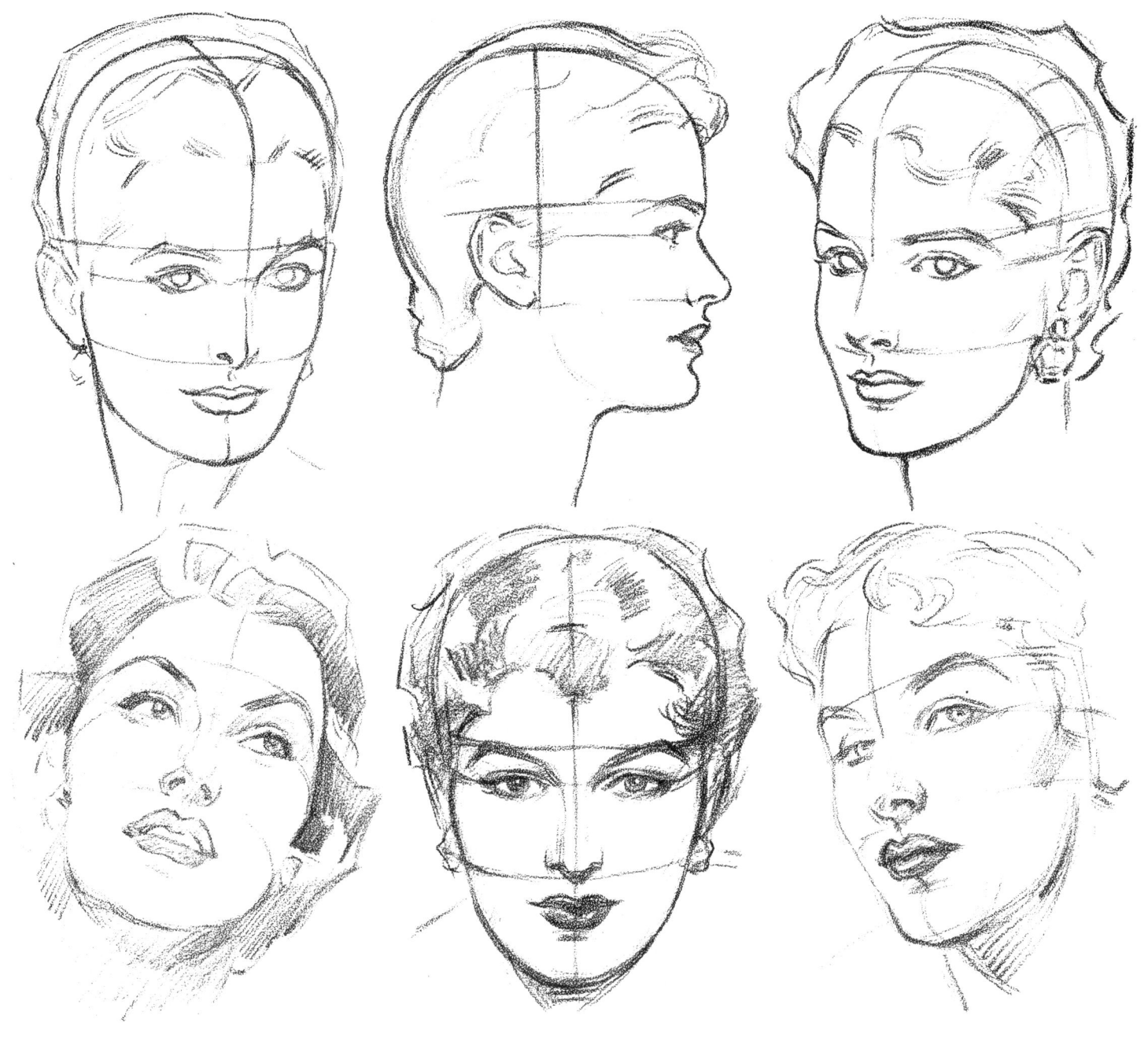

PLATE 41. Establish the construction of each head

It is almost impossible to draw a beautiful woman unless the construction and placement of features are accurate. Keep the nostrils small and watch carefully the placement of the jaw and ears. The eyes and mouth must be in perfect placement and drawing to avoid some very strange and unpleasant results. Just now the brows are left fairly thick. A few years back they were just a thin line. Personally, I like natural-looking brows, but brows and lips, since they are so often made up, follow the trends of fashion. The same is true of hair-dos. Look for the mass effect of forms in the hair rather than the detail. Beauty of face is beauty of proportion, so learn the proportions first; then study your subject individually. The fashion magazines contain quantities of material for study, and will also keep you up to date on make-up and hair styles. Be careful not to draw flat lips. Place the highlight on the lip very accurately; if it is in the wrong place it can change the mouth and the whole expression.

78

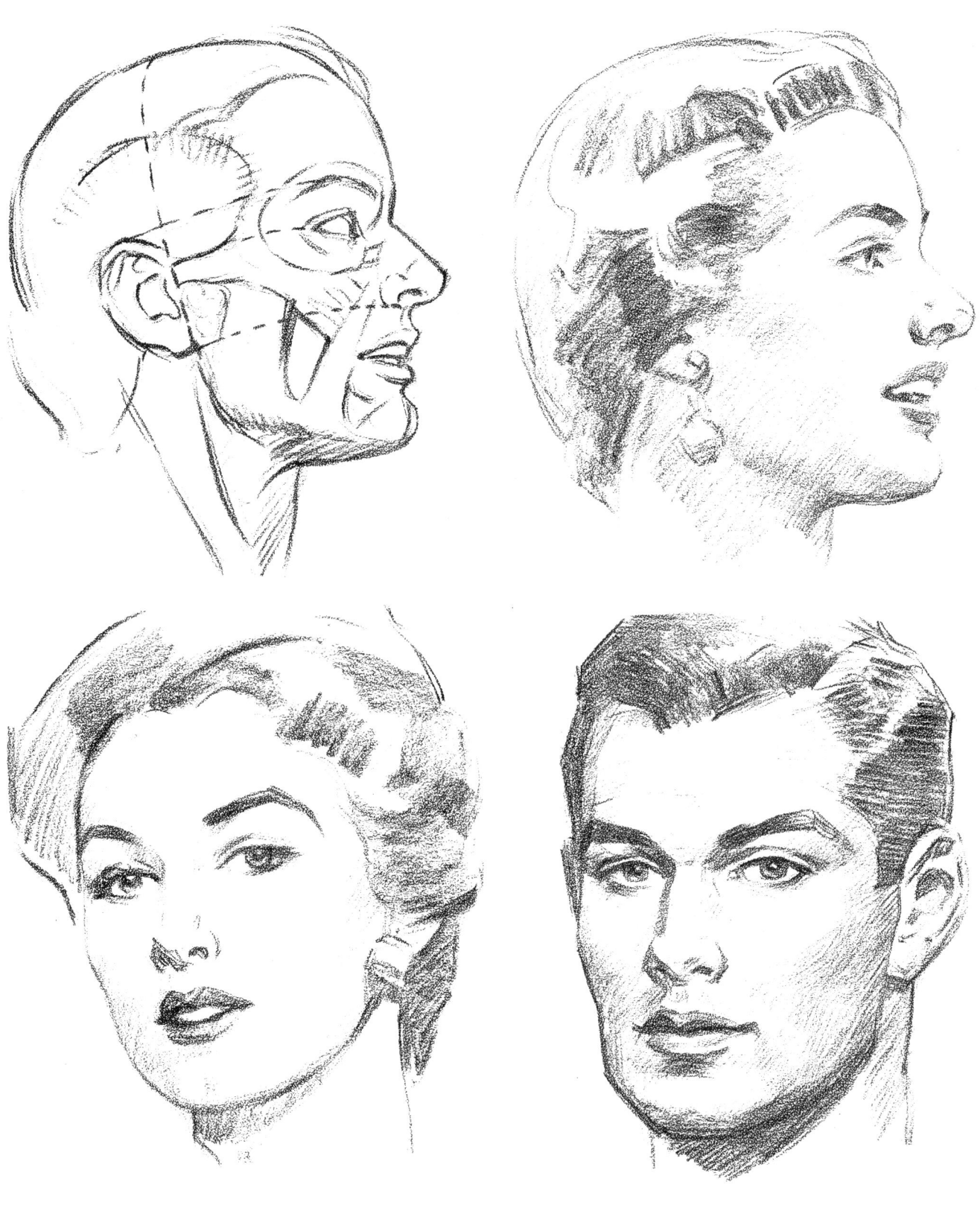

PLATE 42. Bone and muscle are less apparent in women's heads

The underlying anatomy of a girl's head is shown at the top of the page. In drawing a fairly young woman, we let very little of the anatomy show on the surface, though we must know what is underneath to make the surface convincing. At the bottom of the page a male and a female head are shown for direct comparison. Note the heavier bone and muscle construction and the more obvious planes in the male head.

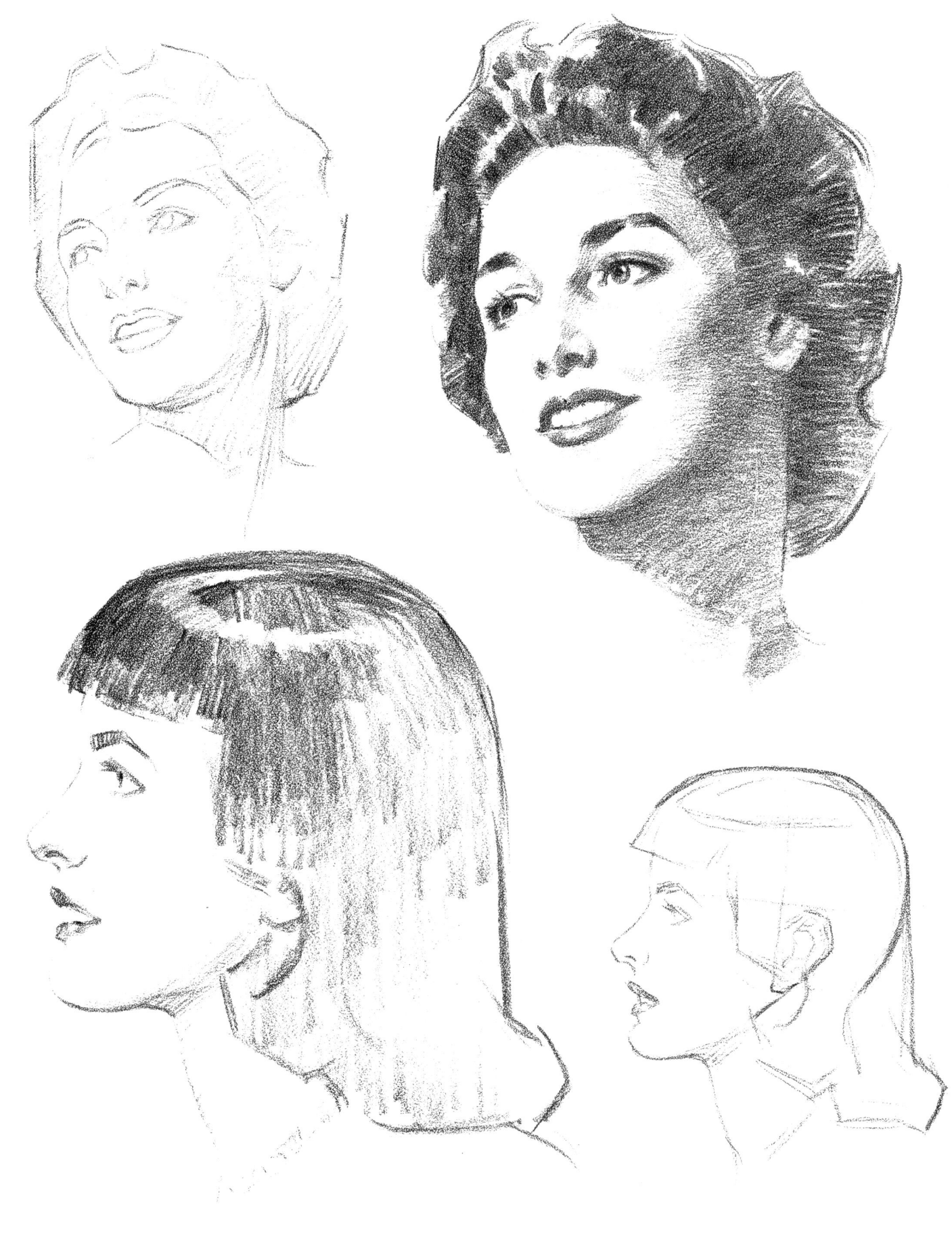

PLATE 43. Charm lies in the basic drawing

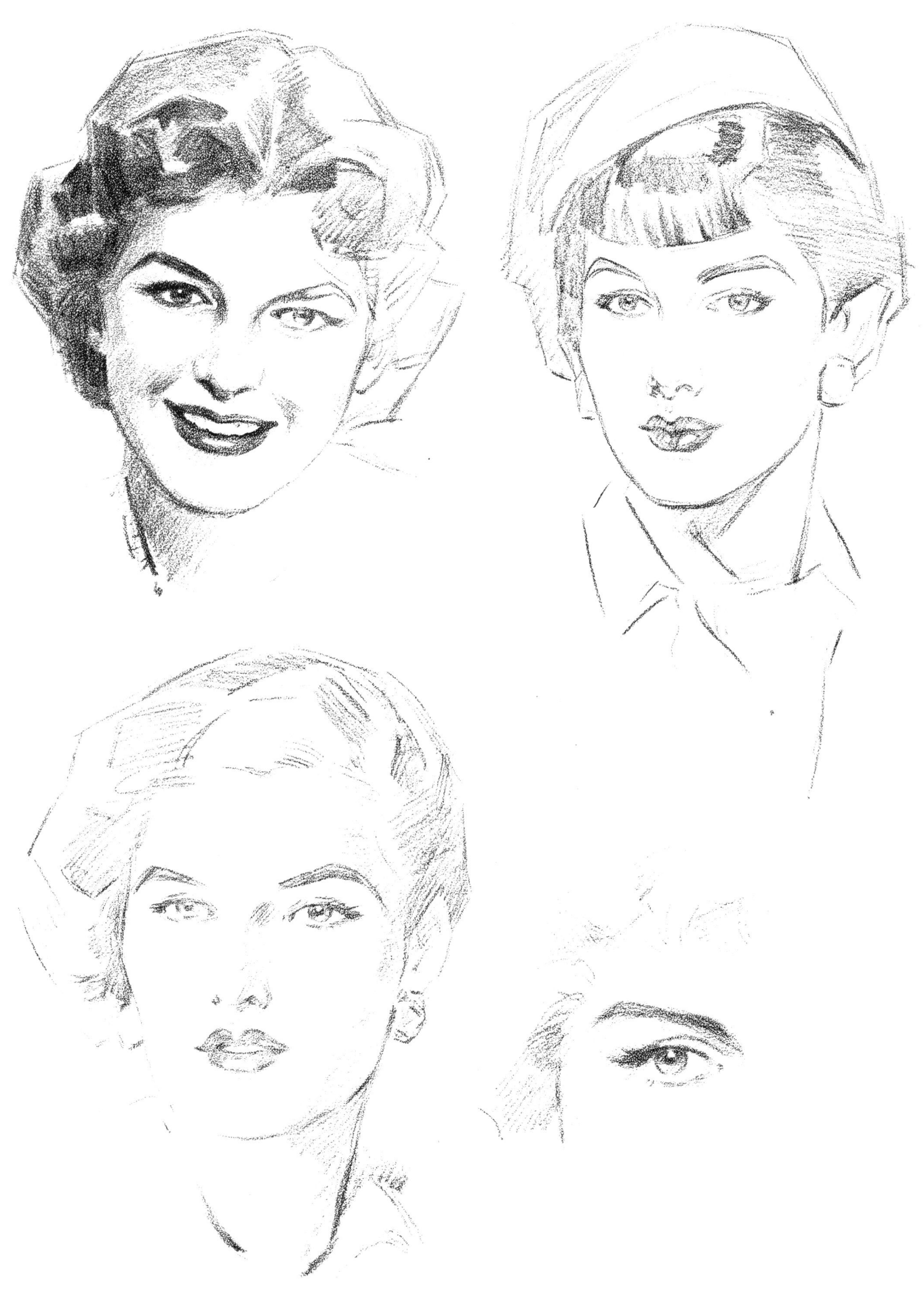

PLATE 44. "Blockiness" also applies to women's heads

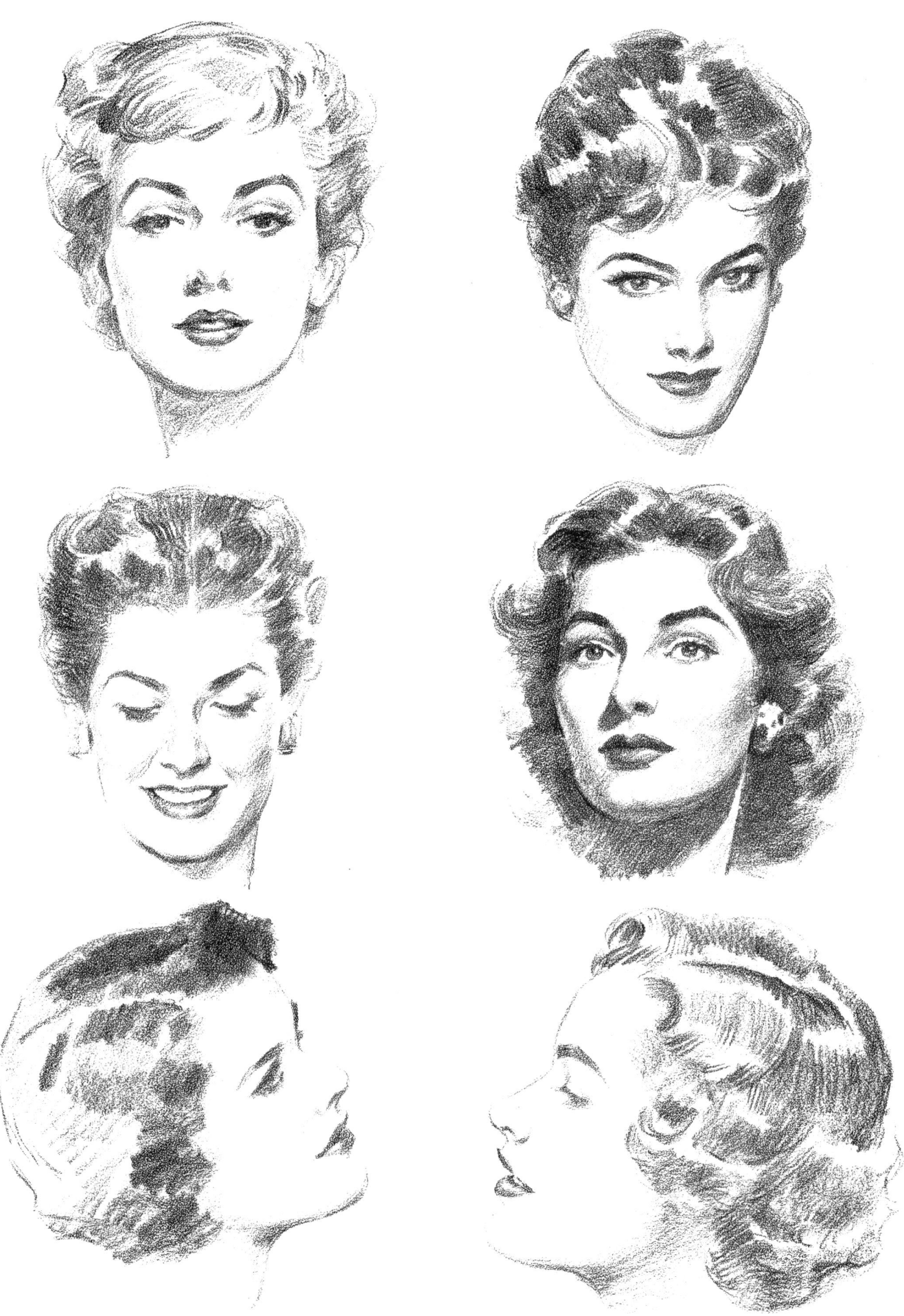

PLATE 45. Some girls' heads

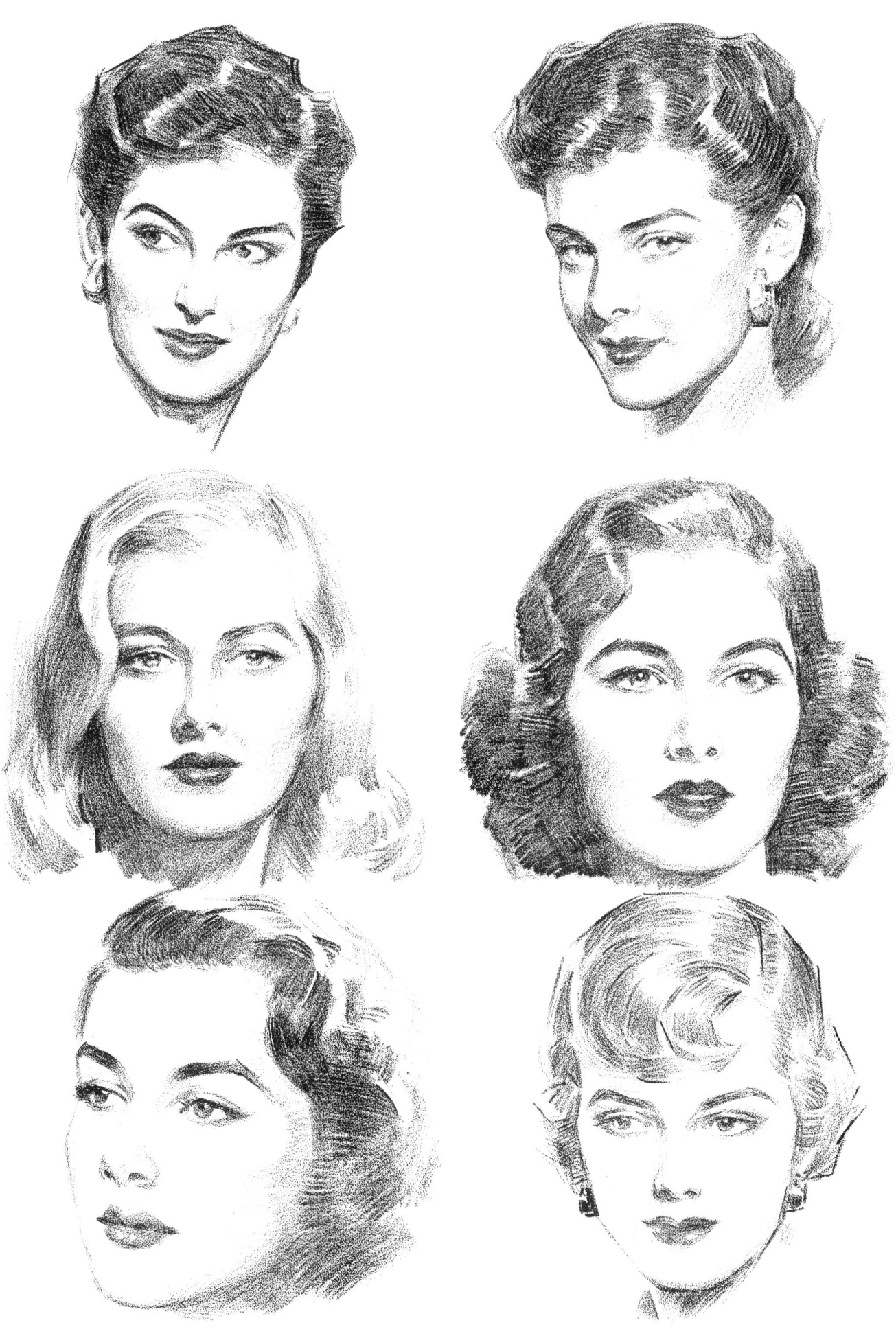

PLATE 46. More girls' heads

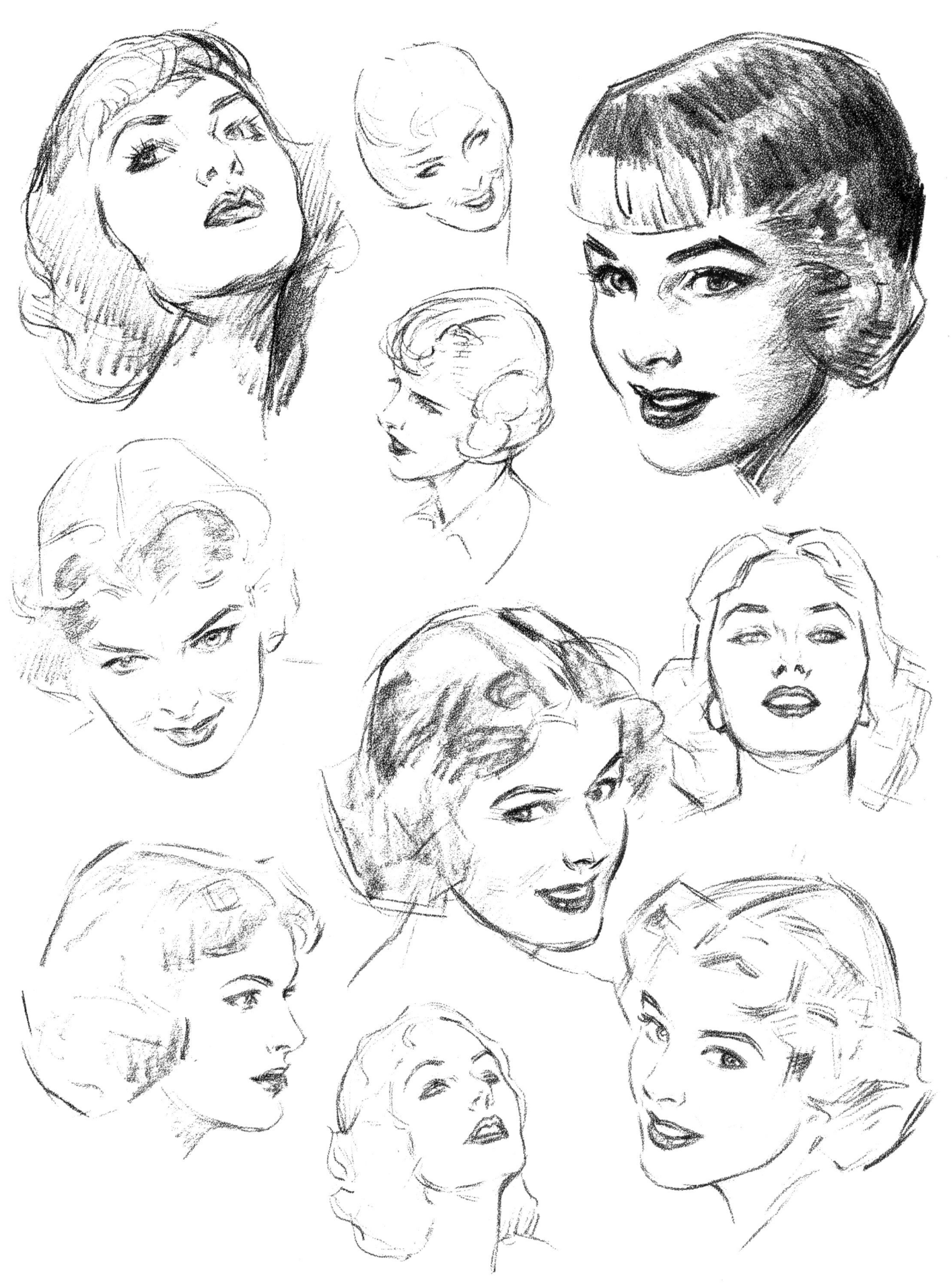

PLATE 47. Sketches

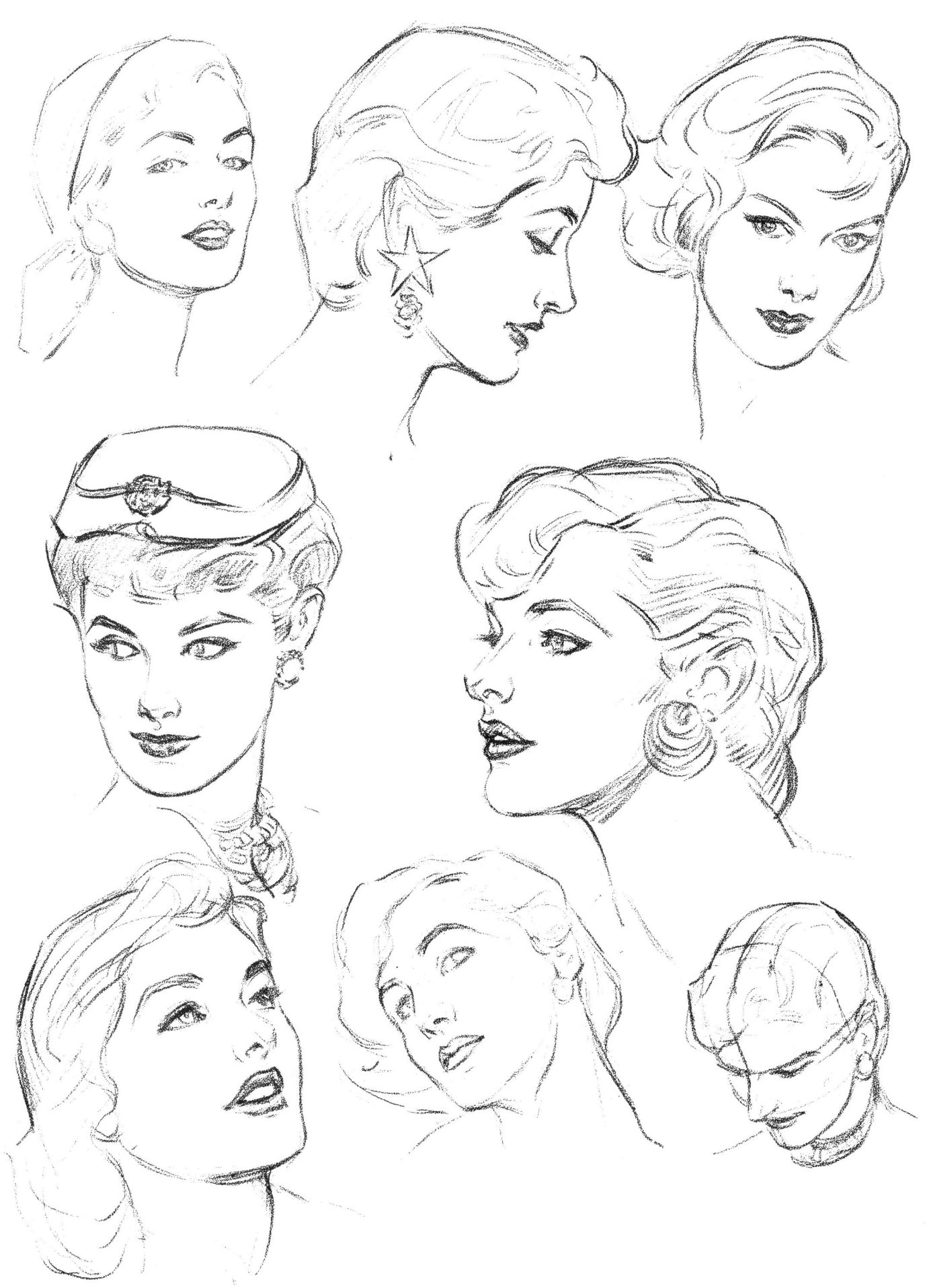

PLATE 48. Sketches

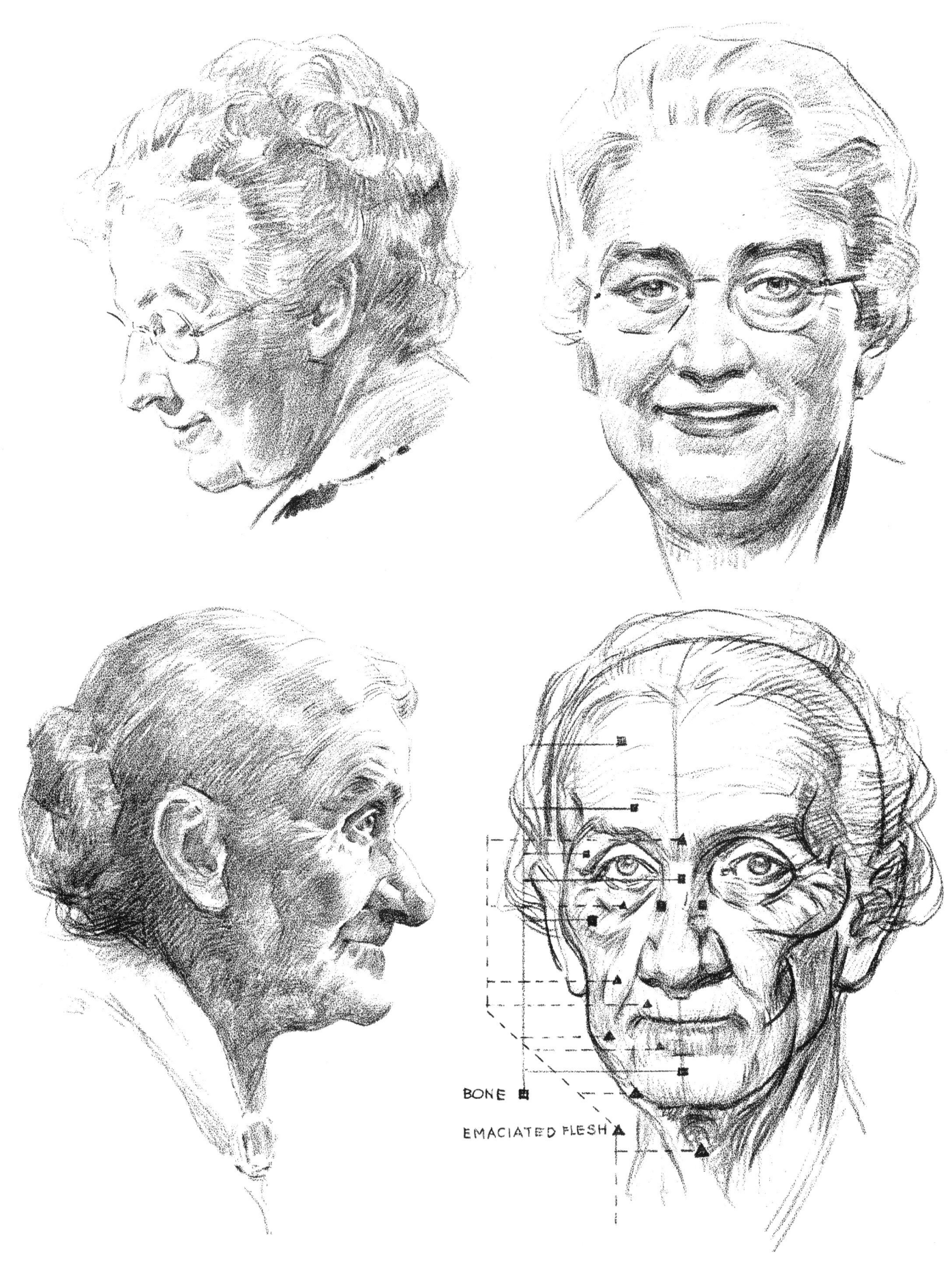

BONE

EMACIATED FLESH

PLATE 49. Grandmothers

86

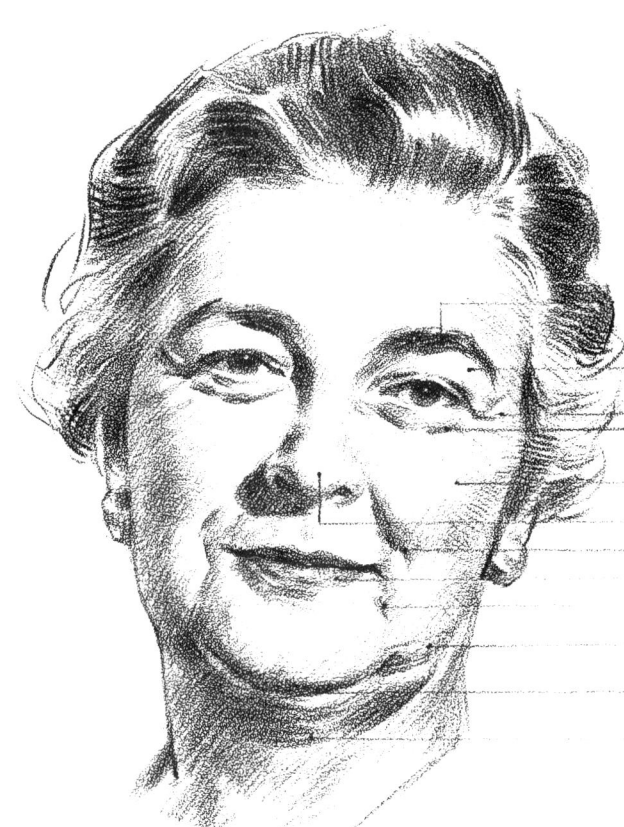

EYEBROWS LOSE SHARP DEFINITION

FLESH DROPS

PERMANENT CREASES FORM

STRUCTURE HERE STILL FIRM

NOSE LOSES SHARP DEFINITION OF CARTILAGE

PERMANENT CREASE FORMS

LIPS LOSE FULLNESS

SLIGHT CREASE FORMS

CHEEK DROPS WITH DEFINITE ACCENT

DEEP CREASE FORMS

FLESH DROOPS UNDER CHIN

HAIR THINS OUT

TEMPLES DEEPEN

BONE OF EYESOCKET VERY EVIDENT

FORMATION OF POUCHES UNDER EYES

CHEEK BONES EVIDENT

NOSE BECOMES BULBOUS

EAR LOBES ENLARGE

CHEEK STRUCTURE LOSES VITALITY

LIPS BECOME THIN LINE

FLESH SAGS BELOW JAWBONE

NECK COMPLETELY EMACIATED

PLATE 50. The aging process

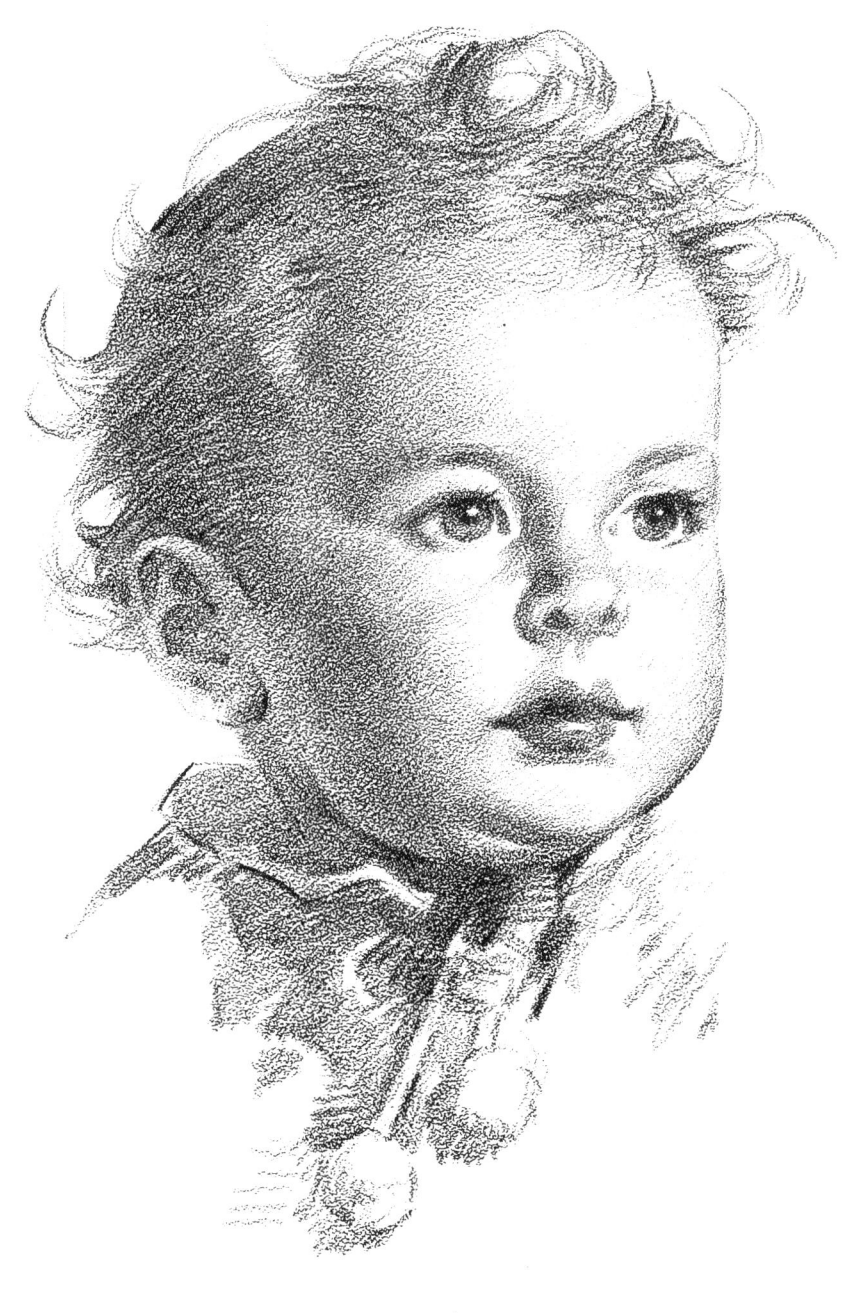

Part Three: Babies' Heads

Part Three: Babies' Heads

DRAWING BABIES is almost a branch of art in itself. Yet the illustrator and commercial artist may be called upon quite often to include them in his work. Babies also make particularly attractive pictures for framing; when they are well done, most families are delighted with them.

If the baby head is understood, it is really no harder to draw than any other head, and sometimes not as hard. The reason is that the artist is dealing much more with construction and proportion than with anatomy. The skull is important, as always, but the muscles are so deeply hidden that they hardly affect the surface. As Plates 51 and 52 show, the proportions are somewhat different from those in the adult head.

In the baby head the bone structure is not yet completely developed. The jawbone, cheekbones, and the bridge of the nose are relatively much smaller. This makes the baby face smaller in proportion to the skull, so that the face, from the brows down, only occupies about one-quarter of the whole area of the head. The cartilages of the nose are way ahead of the bone structure, so the little nose usually turns up, because the bridge above it is rounded and close to the plane of the face. The upper lip is longer, and the chin, being undeveloped, usually recedes or is well under the lips.

Only the iris of the eye is fully developed, which makes the eyes appear large and buttony. They appear to be farther apart than the average adult's eyes because they rest in a smaller head. Eyes set too close together are unpleasant in a baby face and can spoil a drawing. A baby's head can best be studied when the baby is sleeping. Otherwise we must turn to photographs or magazine illustrations. Babies are bound to wriggle and there is nothing that we can do about it. It is therefore of great importance to fix the general or average proportions in your memory.

You will find that a certain blockiness of planes and edges also helps to put vitality into a drawing of a baby. Babies' faces are so smooth and so round that if we copy that quality too meticulously the final effect may lack character.

If you are disturbed by seeing edges of planes in a drawing of a baby face it is probably because you are too close to your drawing. Step back before you change it. Maude Tousey Fangel, one of the greatest baby artists, draws quite vigorously in angles and planes. Mary Cassatt, the Impressionist painter and student of Degas, also had this quality in her work.

Plate 53 shows that the general shape of the baby's head is a bulge attached to a round ball. The distances up and down between the features are relatively short, and the face seems quite wide. The first build-up of the basic shape should have that cute baby look.

In the sketches in Plate 54, the eyes rest in the lower half of the first quarter division. The top line is the line of the brows; the nose rests on the line of the second division; the corners of the lips on the third; and the chin drops slightly below the line of the fourth division.

Plate 58 shows the four divisions for children three to four years old. Note that the brows are a little above the top line, and the nose, eyes, and mouth have been raised above the division lines. These changes make the baby look slightly older. Actually, we have allowed a little more chin and thereby lengthened the face slightly. Plates 55, 56, and 57 show a number of baby heads, all drawn with the foregoing proportions, but differing a little in character as a result of slight differences in the placement of features and the relationship of the face to the skull. Though the proportions vary only slightly, babies' skulls may differ considerably in shape. We find high, low, or elongated skulls in babies as well as in adults.

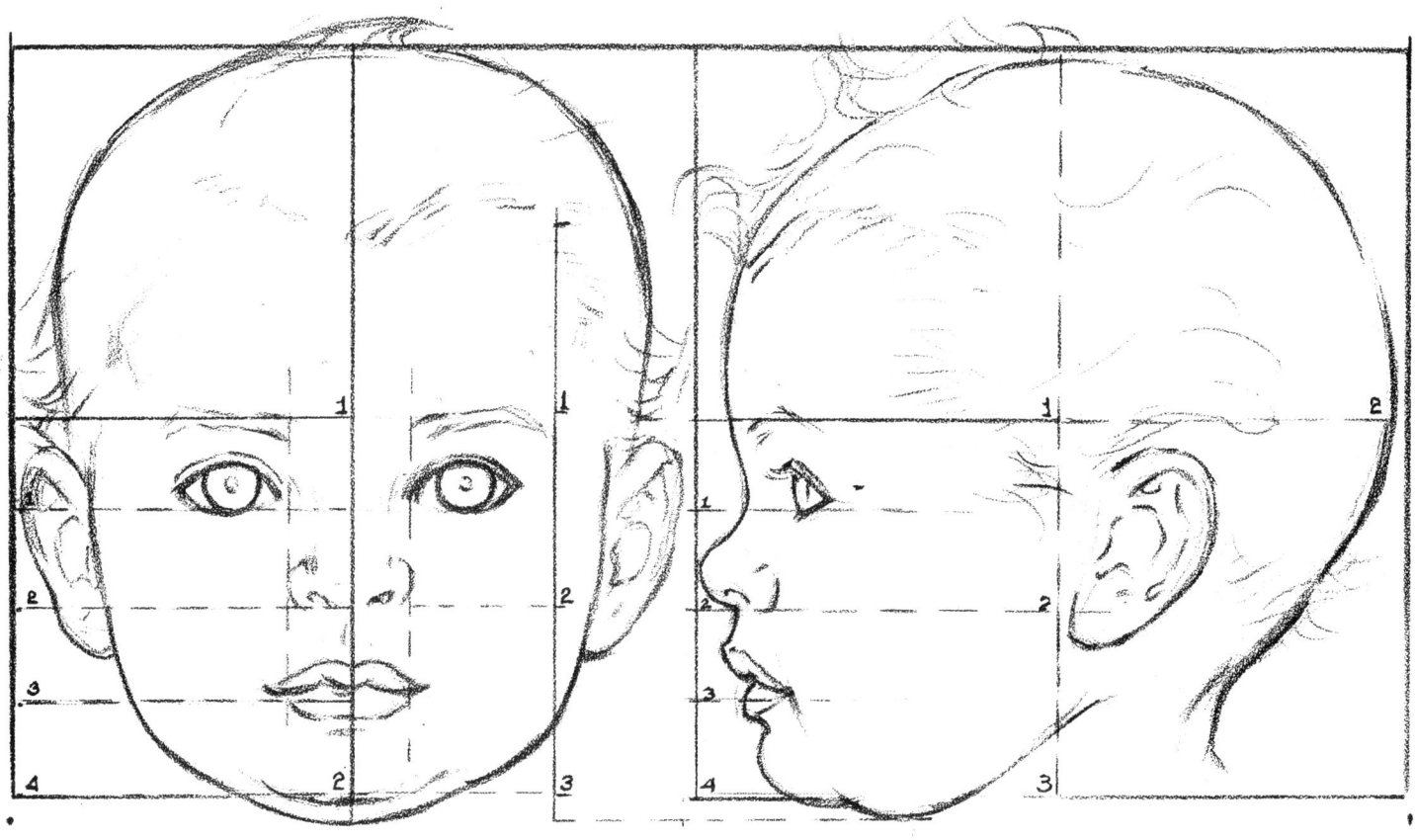

PLATE 51. Proportions of the baby head—first year

Changes in the infant skull take place very rapidly from the moment of birth through the first year or two. It is in the infant stage that the skull takes shape. The original shape may be due to prenatal pressures and the degree of hardness of the bone. After birth the bone tends to adjust to the conditions imposed upon it, the growth of the brain, the closing of the sections of the skull at the top of the cranium, which nature left open and pliable to facilitate birth. Racial skull types are inherited, but the individual type can be purely a matter of circumstance.

In the baby the cranium is much larger in proportion to the face than it is in the adult. The face to the brows occupies about one-fourth of the whole head. This sets the eyes below the halfway point. The most convenient way to set up the baby face is in quarter points. The nose, the corners of the mouth, and the chin come much closer to falling on these points.

As the baby head develops, the face gets longer in proportion to the cranium, which has the effect of moving the eyes and brows upward in the head. Actually, the development of the lower jaw brings that downward, and the nose and upper jaw also lengthen. As a result of these changes the eyes of an adult, and even of a teen-ager, are on the middle line of the head. It is most important to know this, because the setting of the eyes in relation to the middle line across the face is the direct way to establish the age of a child. The iris is fully developed in the baby, and will never get any larger; consequently the eyes look much smaller in the adult face. However, the opening between the eyelids does widen, so that we see more of the eyeball in an adult than we do in a baby.

92

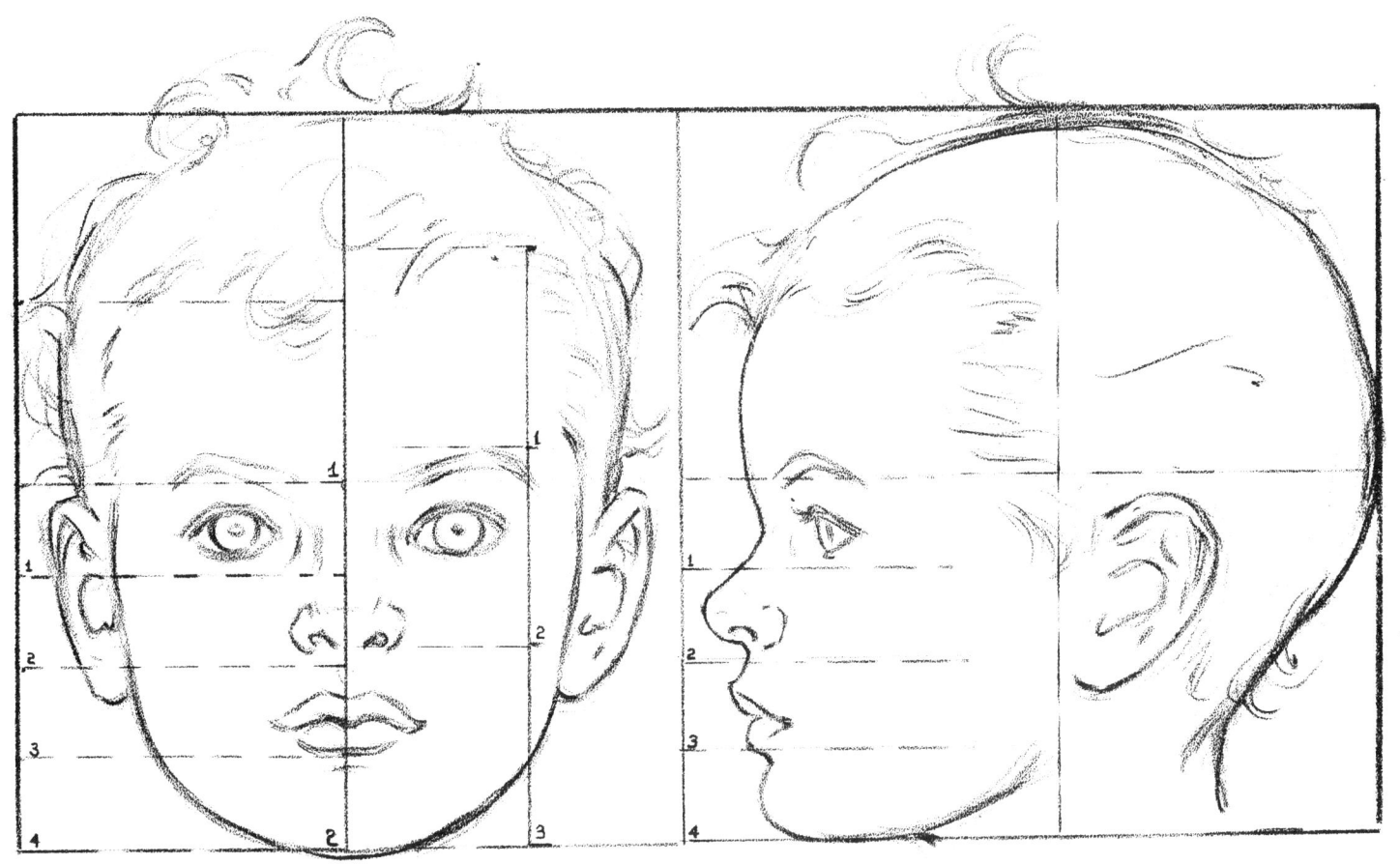

PLATE 52. Proportions of the baby head—second and third years

By the second and third year the eyes are about halfway up the top quarter space, which I have designated the number 1 space. The nose and mouth also appear to have moved up, and the brows now appear to be above the halfway line. Now the lips just touch the bottom of the third space. The ear has not reached the halfway line. However, the face has reached the proportions of three spaces: hairline to brow, brow to bottom of nose, bottom of nose to bottom of chin. Actually these three spaces are still condensed, and each will grow further. But they maintain their proportions to one another while growing. The ear is still well below the middle crossline. Note the line divided into thirds in the right half of the first drawing.

When drawing babies and children it seems easier to maintain four divisions than to use the three divisions of an adult face. While the actual head is much smaller, the spaces between the features are proportionately wider. The eyes are wider apart; the upper lip is longer; the space from eye to ear appears very wide. You have to struggle with these proportions in order to make a baby look like a baby and not like a little old bald man. The baby mouth is more pursed when relaxed. The upper lip rises sharply to its peak and usually protrudes. The chin is small and well under, with often a little fat under it. Babies' ears vary a great deal, some being quite small and others quite large. They are usually rounder and appear thicker in comparison to the face. Babies' brows are usually light and thin or even quite transparent. They are usually much more evident in dark-haired children. The nose is usually small and upturned, and quite rounded. The bridge of the nose is fairly round since it has not had time to develop. The cheeks are extended and full.

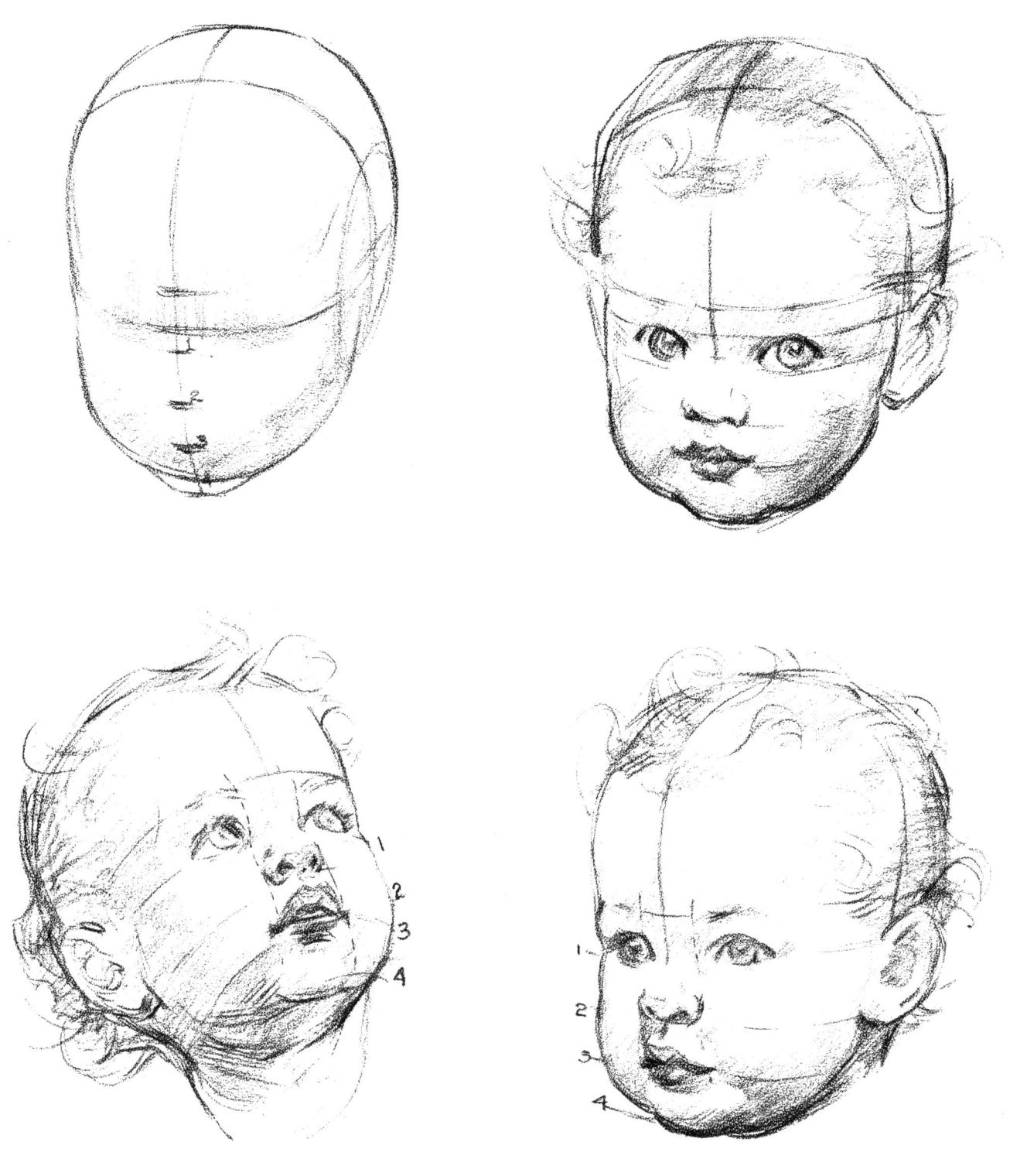

PLATE 53. Construction of the baby head

In drawing a very young baby, draw the ball and plane with the facial plane much shorter. Put the brows on the halfway line. Divide the face from the brows down into four parts. The eyes touch the bottom line of the top division. The nose touches the bottom line of the second division. The corners of the mouth fall on the bottom line of the third division, and the chin drops slightly below the fourth or bottom division. The ear is under the halfway line.

94

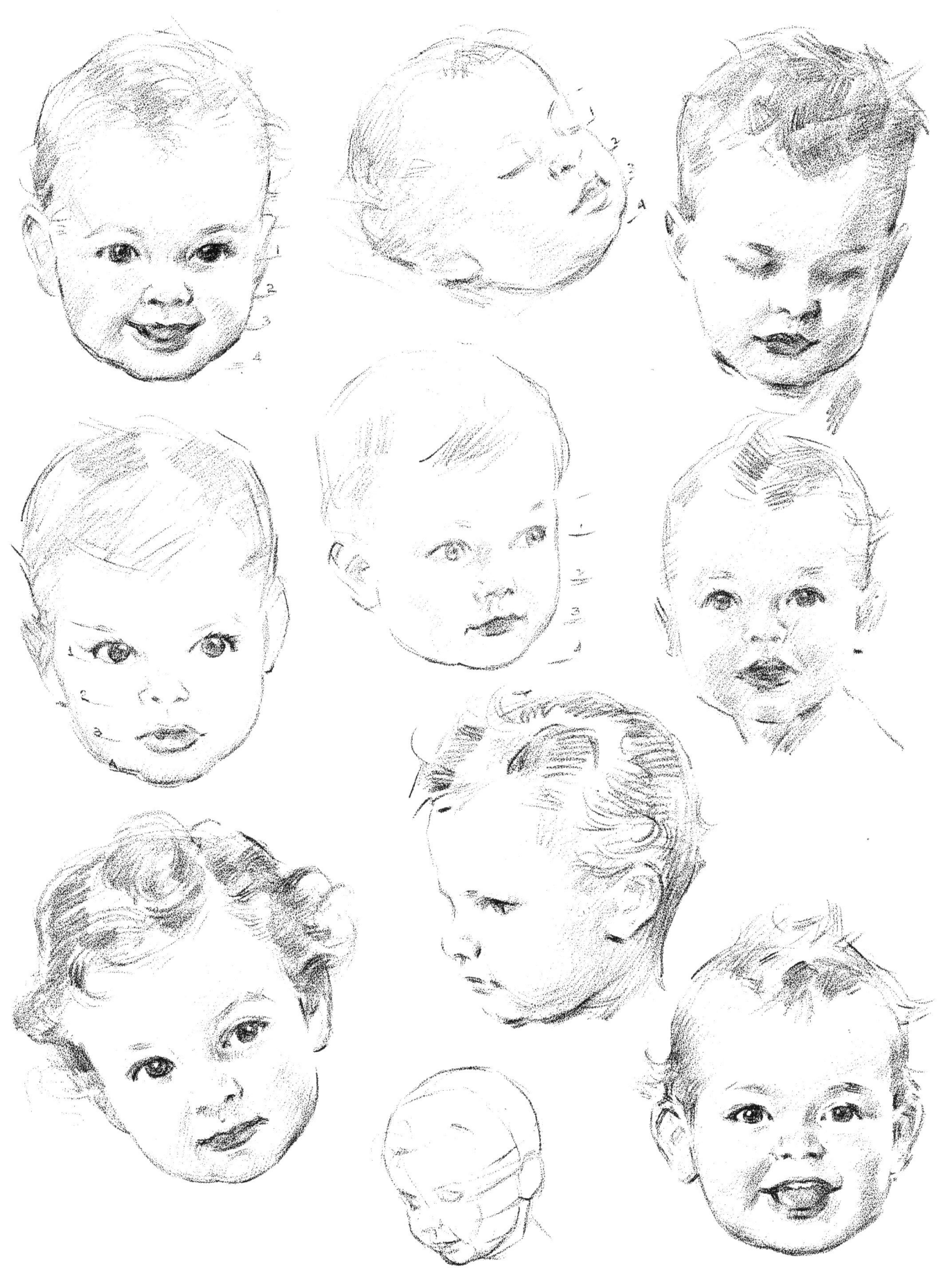

PLATE 54. Sketches of babies

95

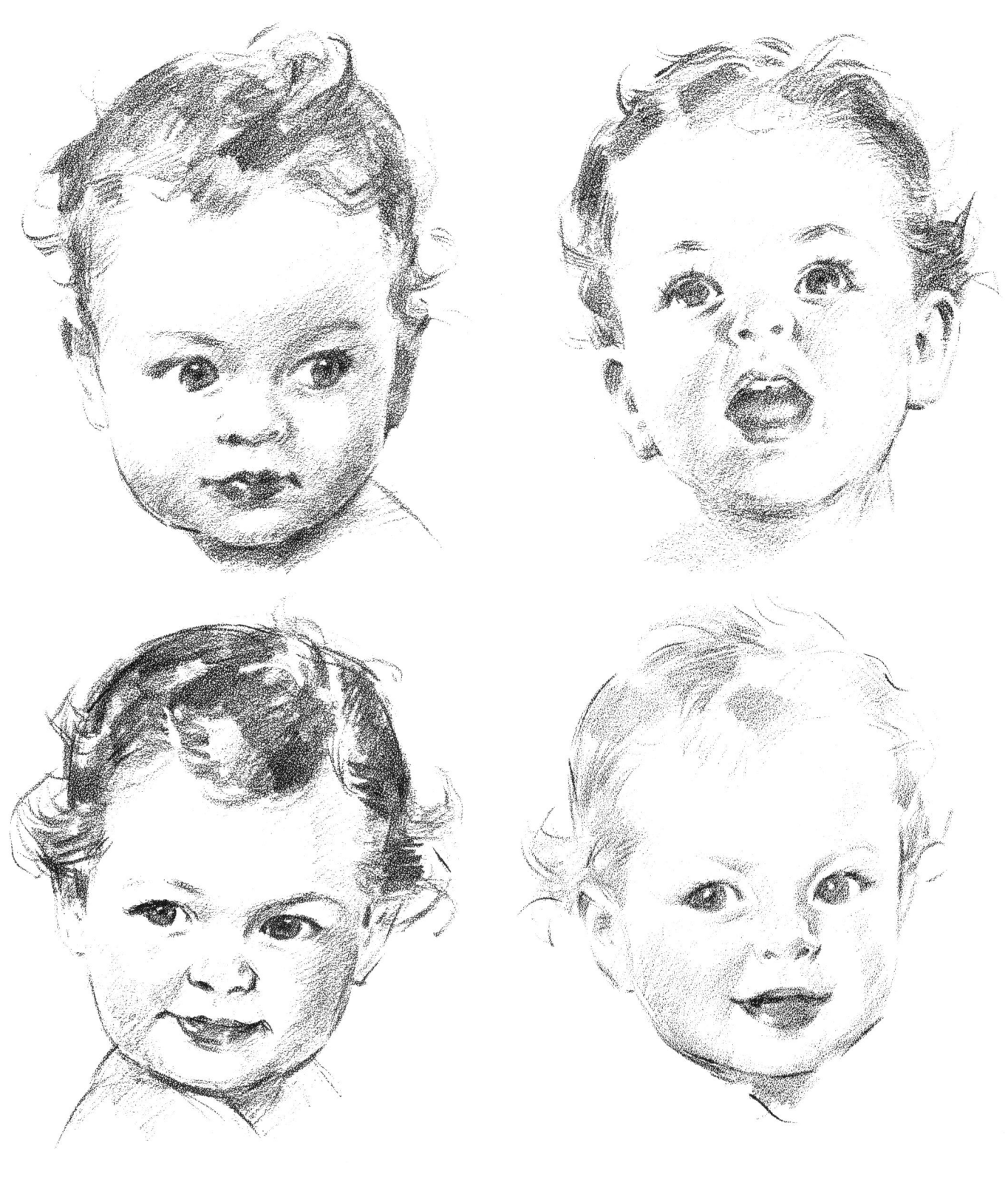

PLATE 55. Studies of babies

The magazines are full of baby pictures, and these are best to practice from, since no baby will hold still long enough for anyone who is not thoroughly familiar with baby proportions to draw from life. The best one can do is to make fast sketches. For this reason finished pictures of babies are usually drawn from photographs, as are the ones on this page.

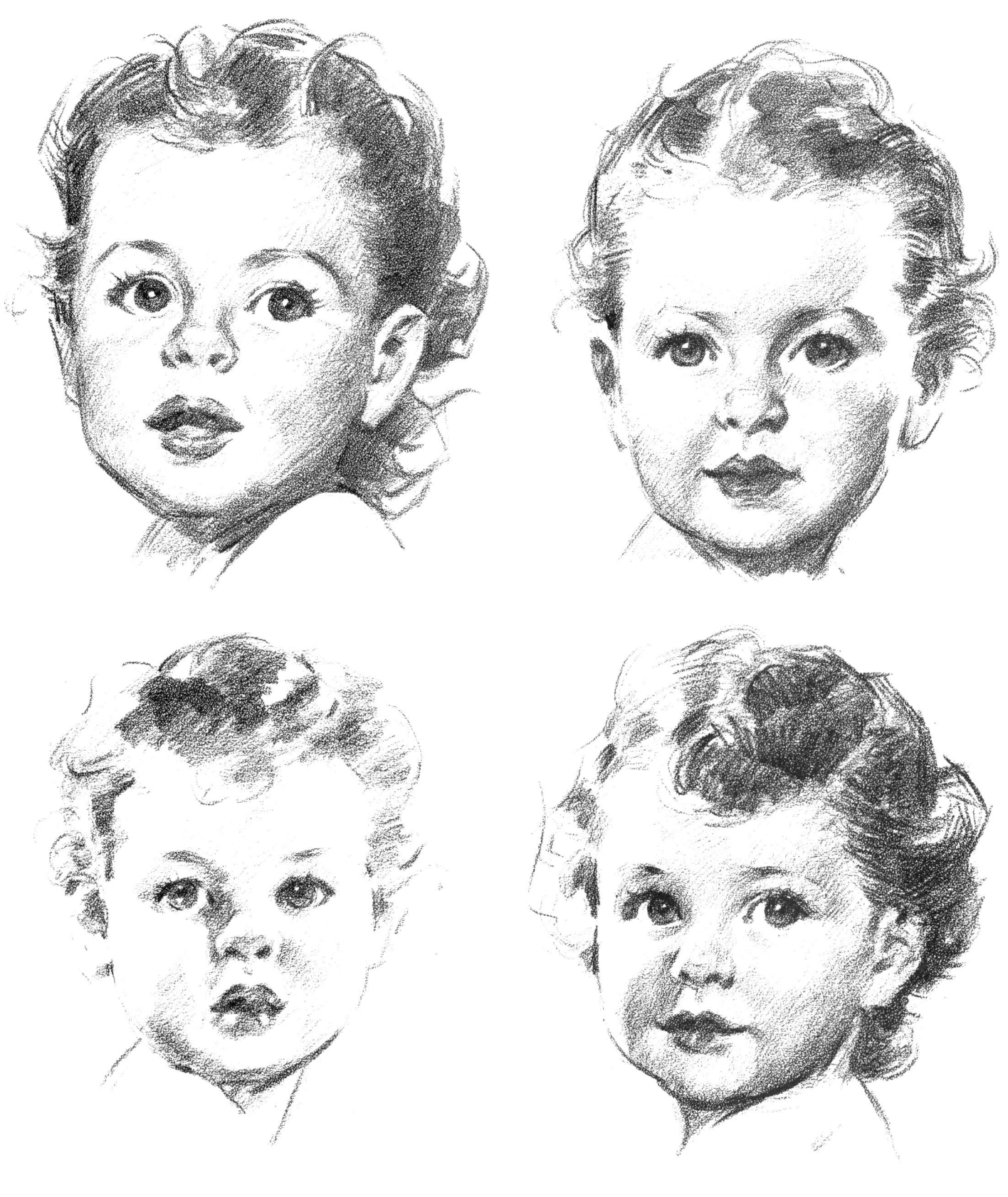

PLATE 56. More studies of babies

As babies grow more hair, they look older, although the proportions have changed only slightly. Some babies develop long eyelashes, which, with their already large and widely spaced eyes, give a great deal of appeal. Go easy on the eyebrows; keep them delicate.

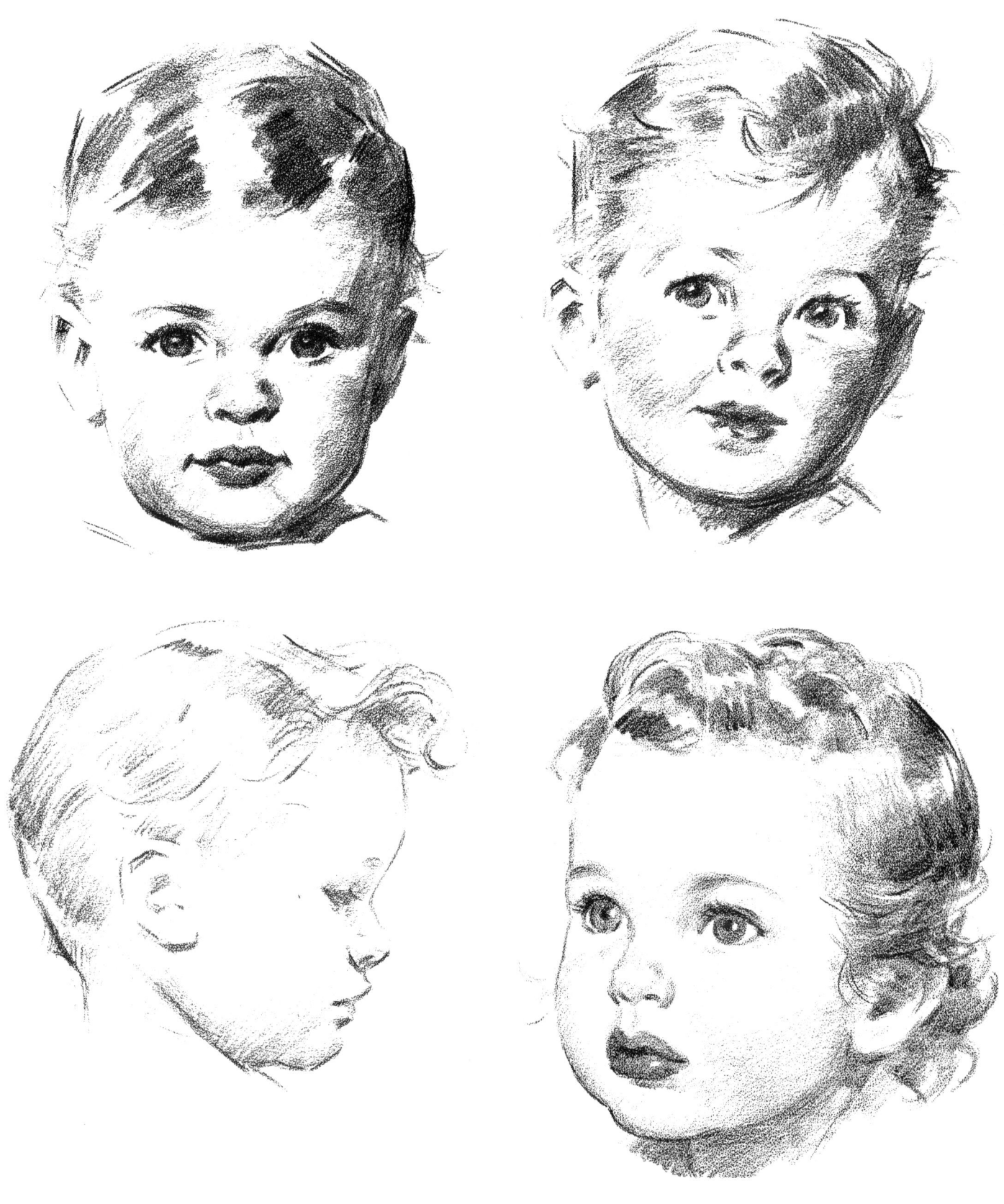

PLATE 57. Some more studies of babies

Remember to keep the bridge of the nose low and concave and the two little round nostrils rather widely spaced. Let the upper lip protrude when the baby is not smiling. Set the ears fairly low, and the chin round and well under. Keep the cheeks high and full. You will usually want to add light tone with a highlight.

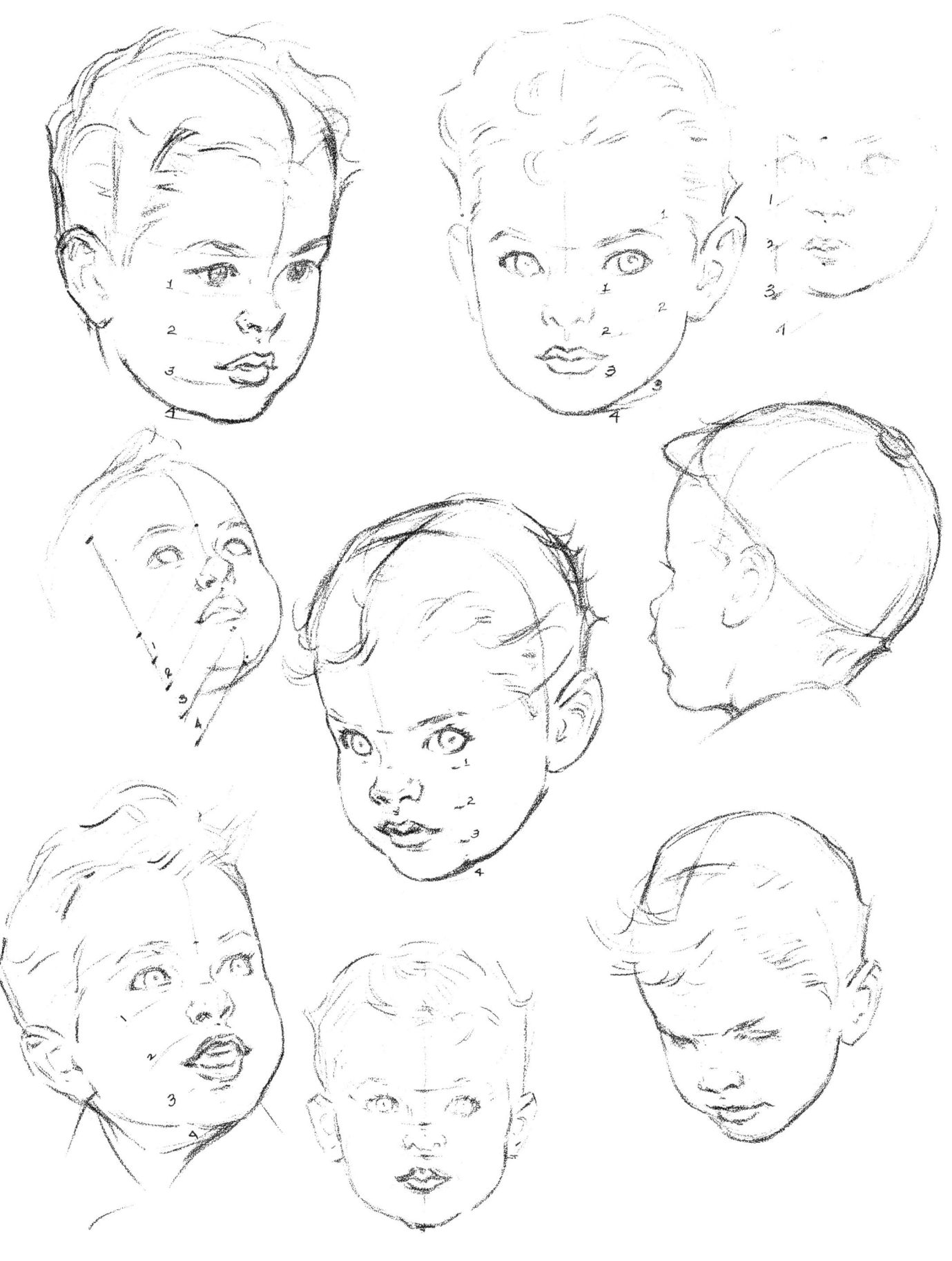

PLATE 58. The four divisions of the face—third and fourth years

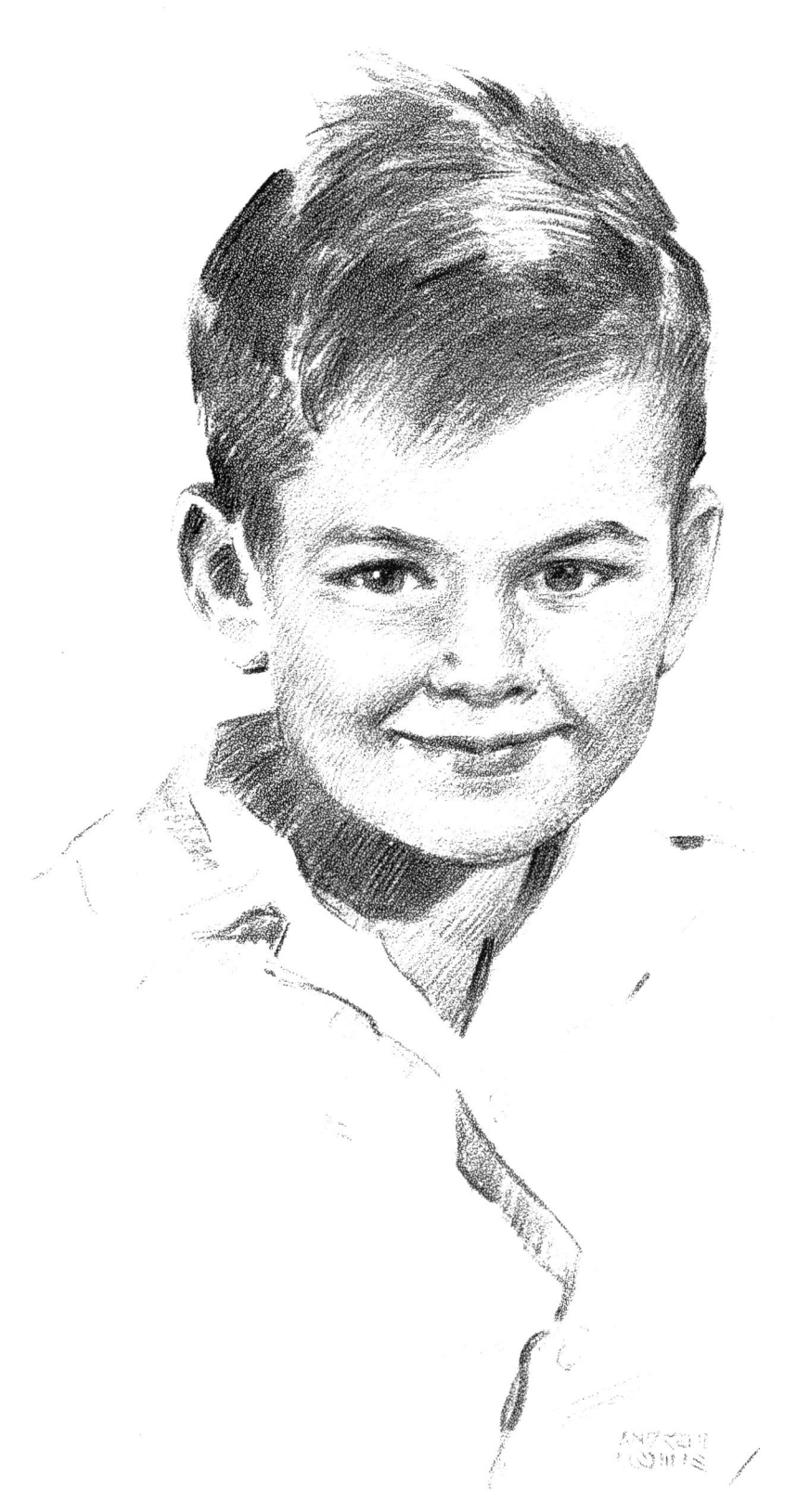

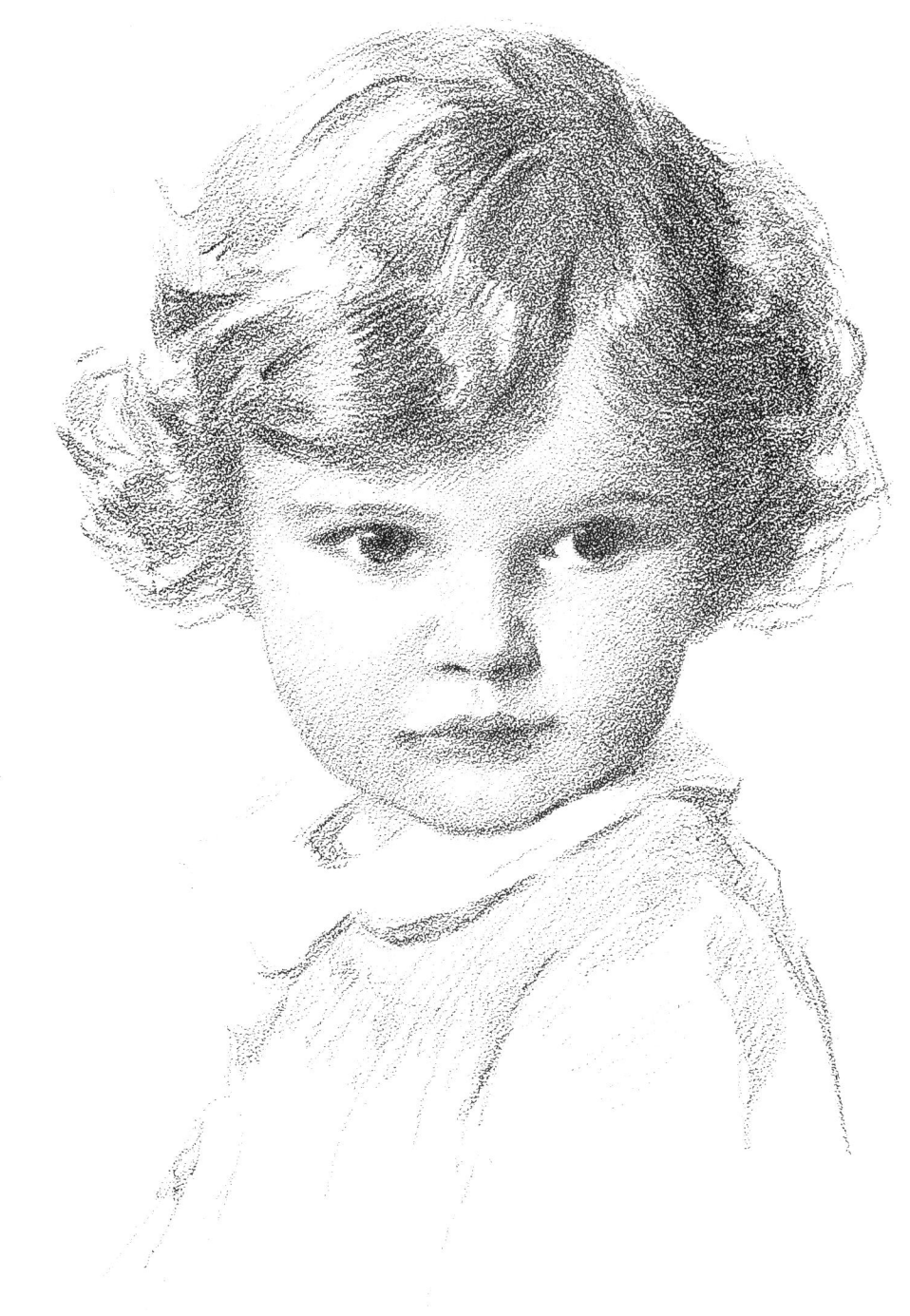

Part Four: Heads of Boys and Girls

I. SMALL CHILDREN

Part Four: Heads of Boys and Girls

I. SMALL CHILDREN

LET US UNDERSTAND that no branch of art can be reduced completely to a formula without endangering the very art that must go into it. We do, of course, seek ways and means to an end, and that end is correctness. Art, however, is not the justification of correctness. Art is not always perfection. Let us say that art is truly a form of expression, and full expression cannot be limited by formula, but only guided toward greater meaning and truth. African sculpture has expression and because of that it is art. It is certainly not truth as we know it, but it may be truth with a greater meaning as they know it. We may reach greater truth by simplification and even by subordinating minor truth. Detail may be minor truth but without real significance. Each hair in an eyebrow is detail and minor truth, but carries little significance. Each blade of grass is detail, but we may be more interested in the whole hillside and the effect of sunlight on it.

In drawing children, let us be guided as much by our feeling toward them as by rules of construction and anatomy. The light on a child's hair may be just as beautiful and intriguing as the light on the hillside. The glint of mischief in the eye of a young boy may really be what we are drawing, more than the perfect anatomical construction of that eye.

It is easy to become so absorbed in technicalities that we miss the purpose. The technical must be united to the spiritual, because technique without spirit is meaningless. But feeling cannot be conveyed without technique and the knowledge behind technique.

Every area of every drawing, painting, or any other expression of form should be a part of a whole design. The lights and shadows, the edges, the textures and materials may all be considered as much from the standpoint of design and arrangement as for any other quality. In drawing heads, the pattern of the hair, the shadows cast from the head, and the bit of clothing all offer opportunity for design. The lights and shadows on the face itself create design, good or bad, whether we are conscious of it or not. The whole head is a design of forms fitted together, and it is a masterpiece of design, functionally as well as artistically.

I speak of all this so that we may approach our subject with humility and appreciation of its wonders. To me there is nothing more beautiful or wonderful in the world than the head of a small child. Life has left no scars, no lines of anxiety and frustration; it is the new flower emanating from the bud, fresh and as yet almost untouched.

If children do not move you, it is perhaps a mistake to try to draw them. You cannot draw them effectively from too great an emotional distance. When joy goes out of your work, it is apt to bog down in pure technicality.

It happens that much of my own work has been concerned with drawing children, and the more I do it, the more I find to enjoy in it. I feel that there is a mountain of fascinating truth of which I have barely scratched the surface, and this comes after drawing and painting perhaps thousands of heads of adults. Drawing children has a vast and relatively unexploited commercial market. We need more drawings of children and fewer photographs, both in advertising and on our walls. The fact that children cannot sit still need not discourage you. You can trace from photographs and still raise the quality of your rendering beyond the purely photographic detail to a more artistic expression.

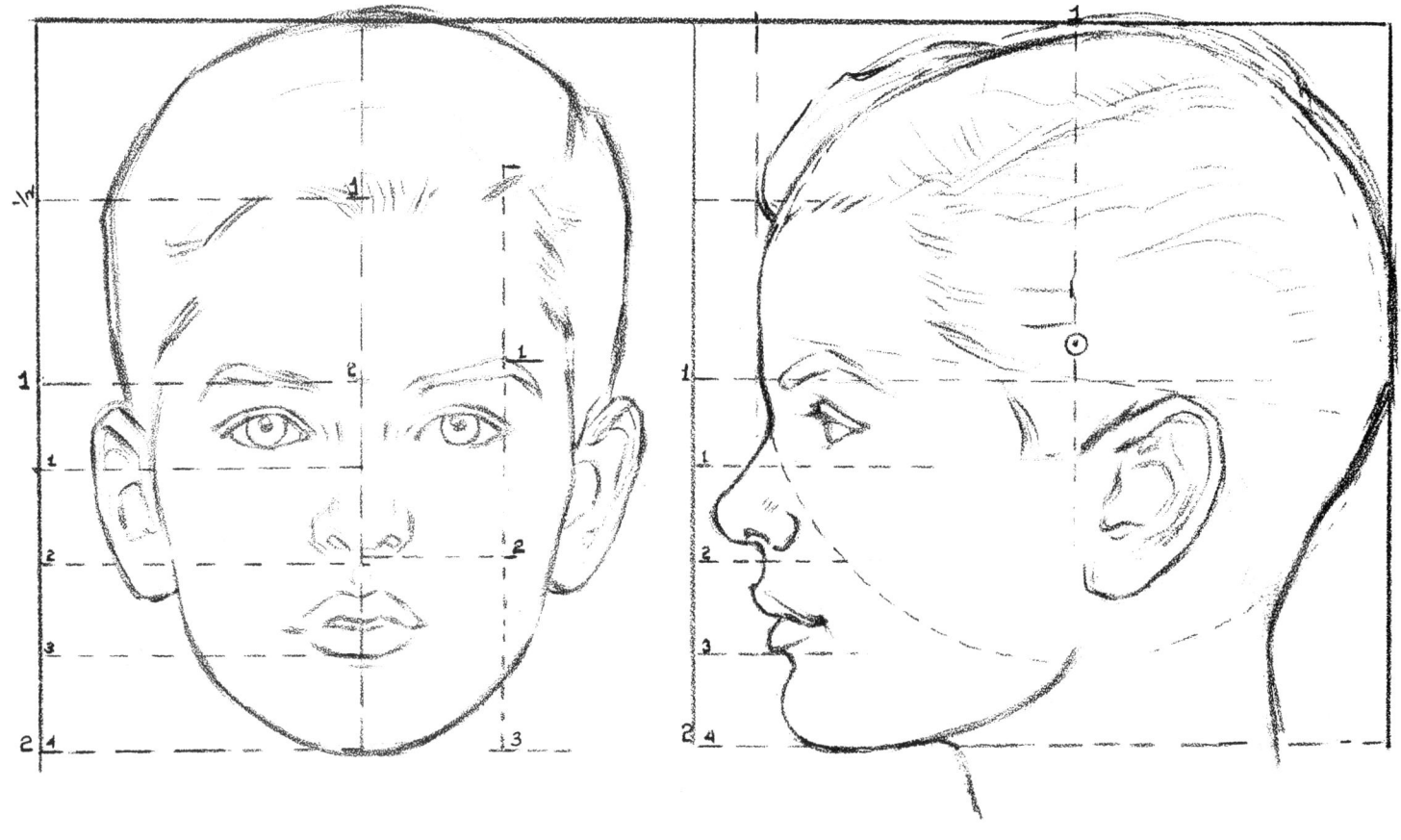

PLATE 59. Proportions of the little boy's head

In the small boy the up-and-down proportions are about the same as those in the older baby. But now the face is relatively narrower, coming well inside the square in the front view. The eyes appear smaller, because they do not grow and the face does. We can only use the large "button" eyes for very young children. The jaw and chin of the boy pictured above have started to grow, making the chin more prominent. The bridge of the nose is higher, and the nose is a little longer, almost touching the bottom of the second quarter. The lips touch the bottom line of the third quarter. At a fairly early age a full shock of hair grows. This accentuates the large cranium but keeps the face looking small and adds to the cuteness of the child. If a child has curly hair, mothers sometimes let the hair grow until it begins to look grotesque. So it is well to know where the cranium really is.

It is hard for little boys to sit still; in drawing them, as in drawing babies, practice from photographs and clippings. Note that the ear is coming up to the halfway line. Little boys' heads seem to extend far back because the neck is small and the muscles which attach to the base of the skull are not yet developed.

Notice particularly that the nostrils have grown and the upper lip appears to be somewhat shorter. The ear grows considerably during this period and the one which follows. I believe the ear is fully developed by the time the child is ten or twelve. The space from the nose to the ear still appears quite wide. Lashes are quite long. The hair grows quite well over the temples.

104

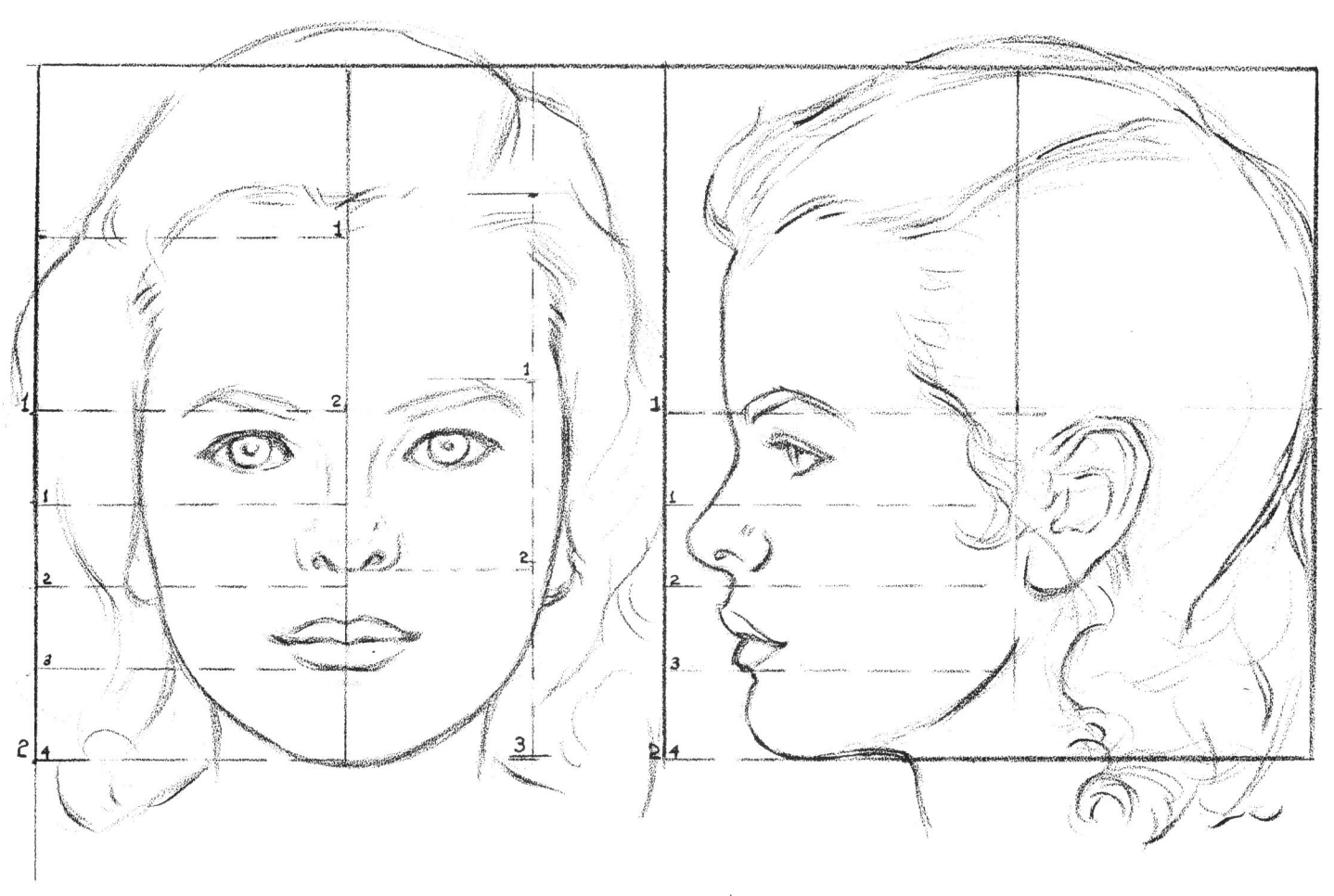

PLATE 60. Proportions of the little girl's head

The proportions of the head are practically the same in little girls as in little boys. Little girls are characteristically wider at the eyes and the jaw and chin are rounder. Very often the crease of the upper lid hardly shows over the eye. All the lines of contour are usually rounder in girls. Knowing this helps you make a little face more feminine; blocky or squarish forms give a little boy a more rugged look. In little girls the forehead tends to be higher at an earlier age than in boys. Some authorities claim that certain qualities of mentality develop faster in girls than in boys. This may account for the higher, wider forehead, I cannot say. I do know that a closer hairline makes a boy look more boyish, while a larger forehead makes a little girl look more girlish. The treatment of the hair helps greatly in drawing little girls.

Care should be taken not to draw the mouth too large on a little girl's face, or too black. This can easily give an adult look, or a theatrical effect not pleasant in children. The little girl's neck is round and small in proportion to the head. The crease between the neck and jaw seldom runs up to the ear but points below it. It is seldom sharply defined. The forehead may easily protrude a little at the top. The planes of the face are all well rounded, but to keep your drawing from looking too smooth and photographic, you can introduce a good deal of blockiness into the hair. The ear is more delicate in structure and it comes up to the halfway line. The brows should also be kept delicate.

105

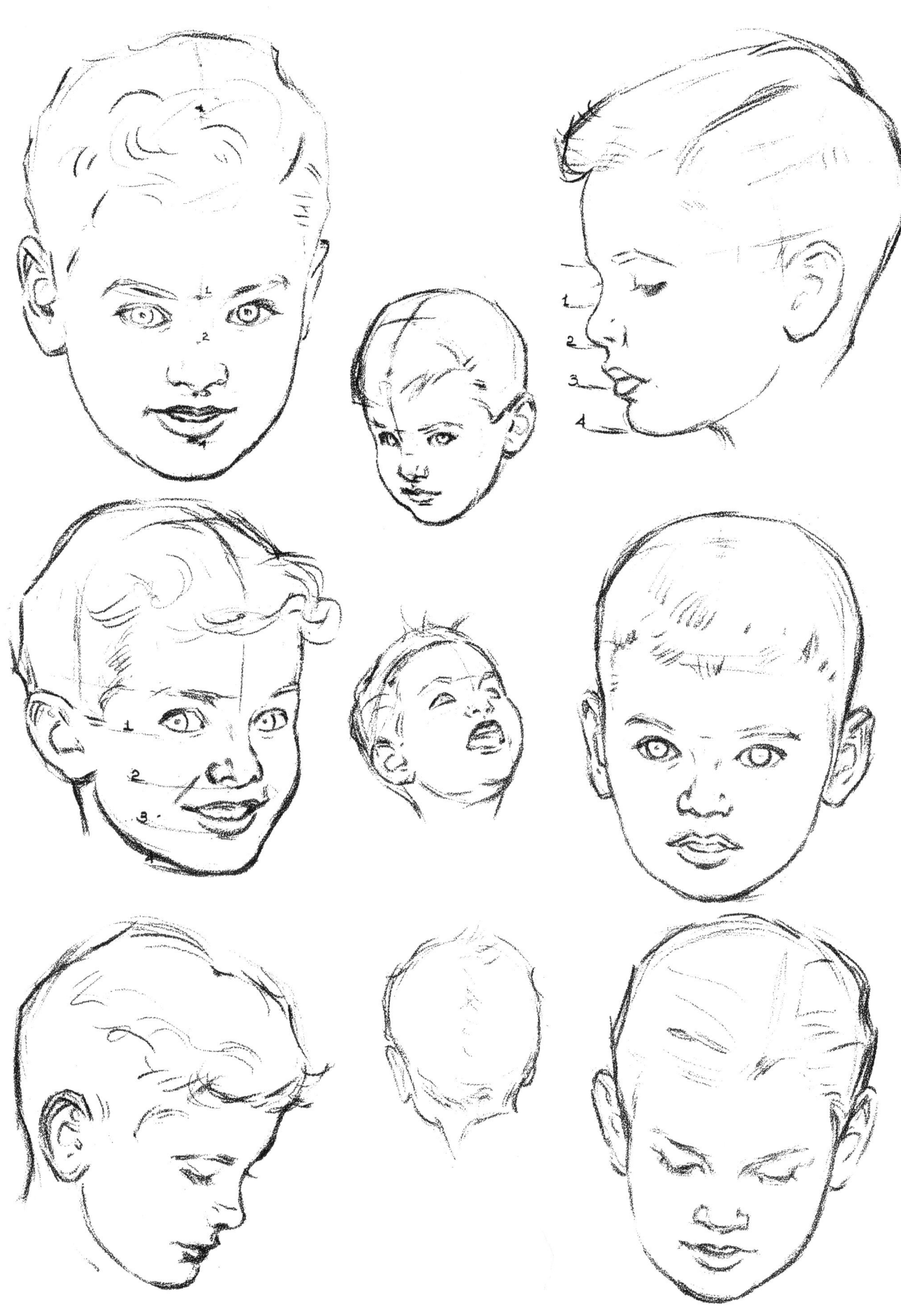

PLATE 61. Construction of the little boy's head

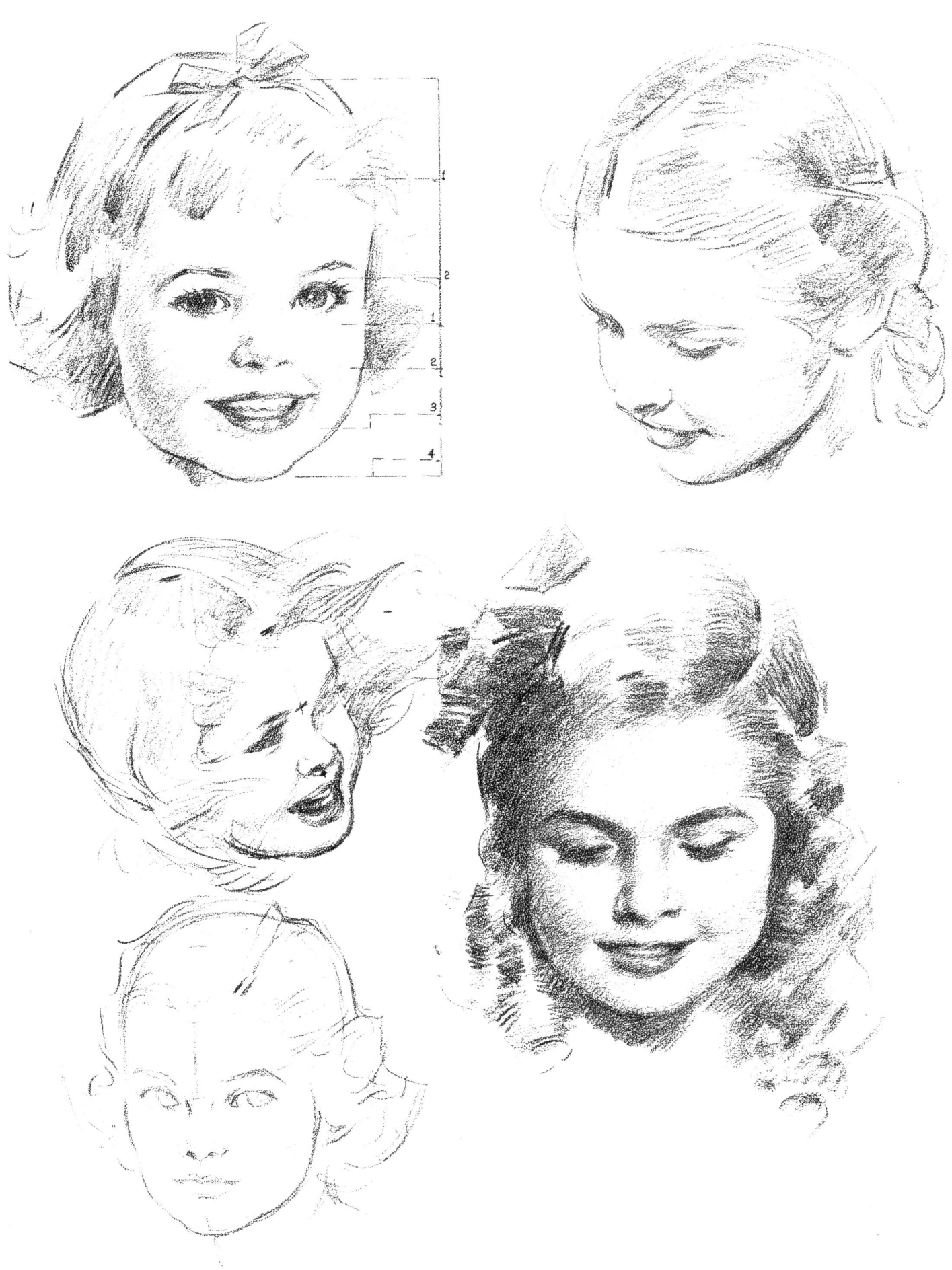

PLATE 62. Construction of the little girl's head

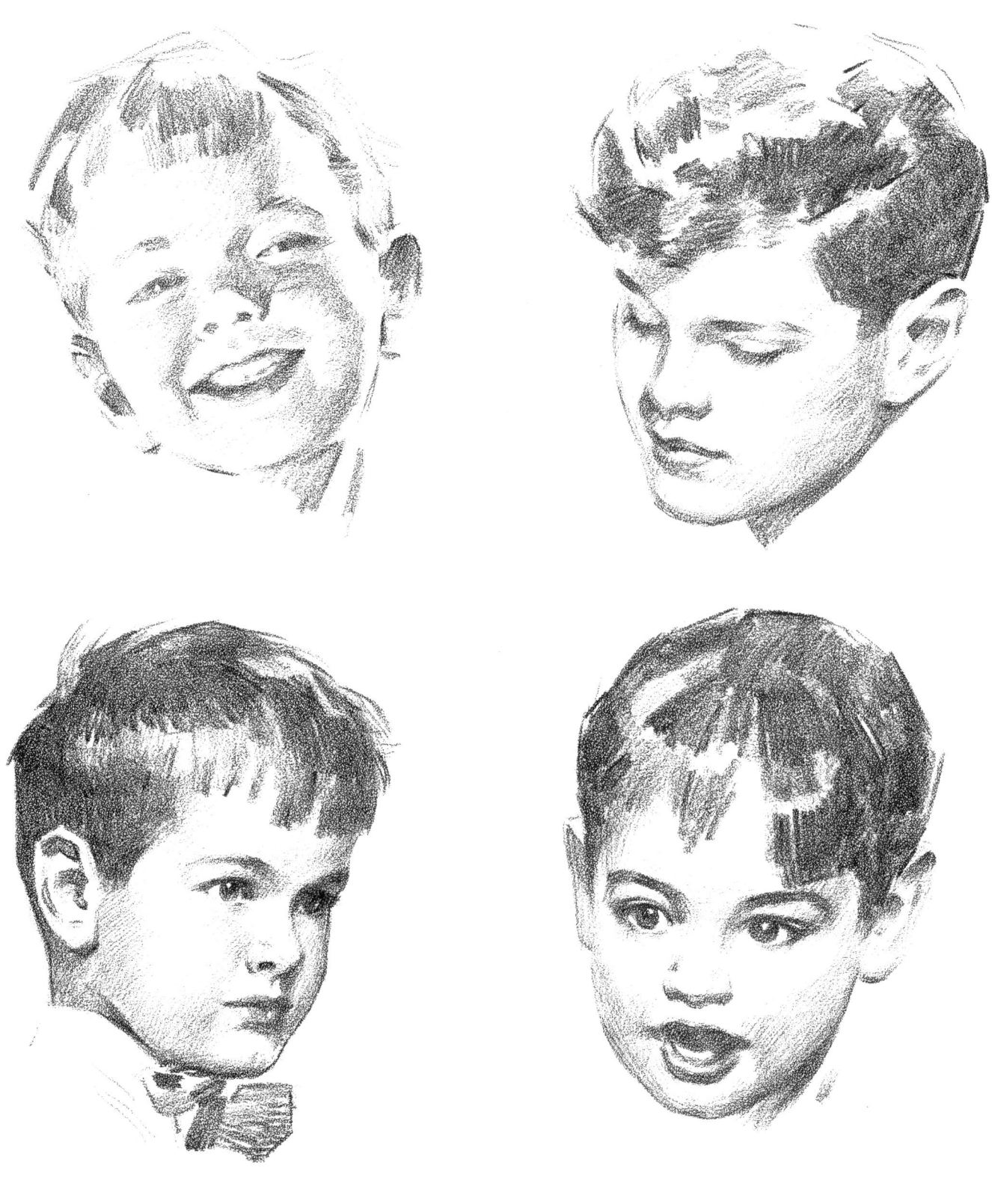

PLATE 63. Studies of little boys

Sometimes back lighting or rear top lighting is effective in combination with front lighting in drawing heads. The important thing is not to allow two lights to fall on the same surface, because this type of lighting cuts the area into crisscross shadows. Build up the hair in blocky forms.

108

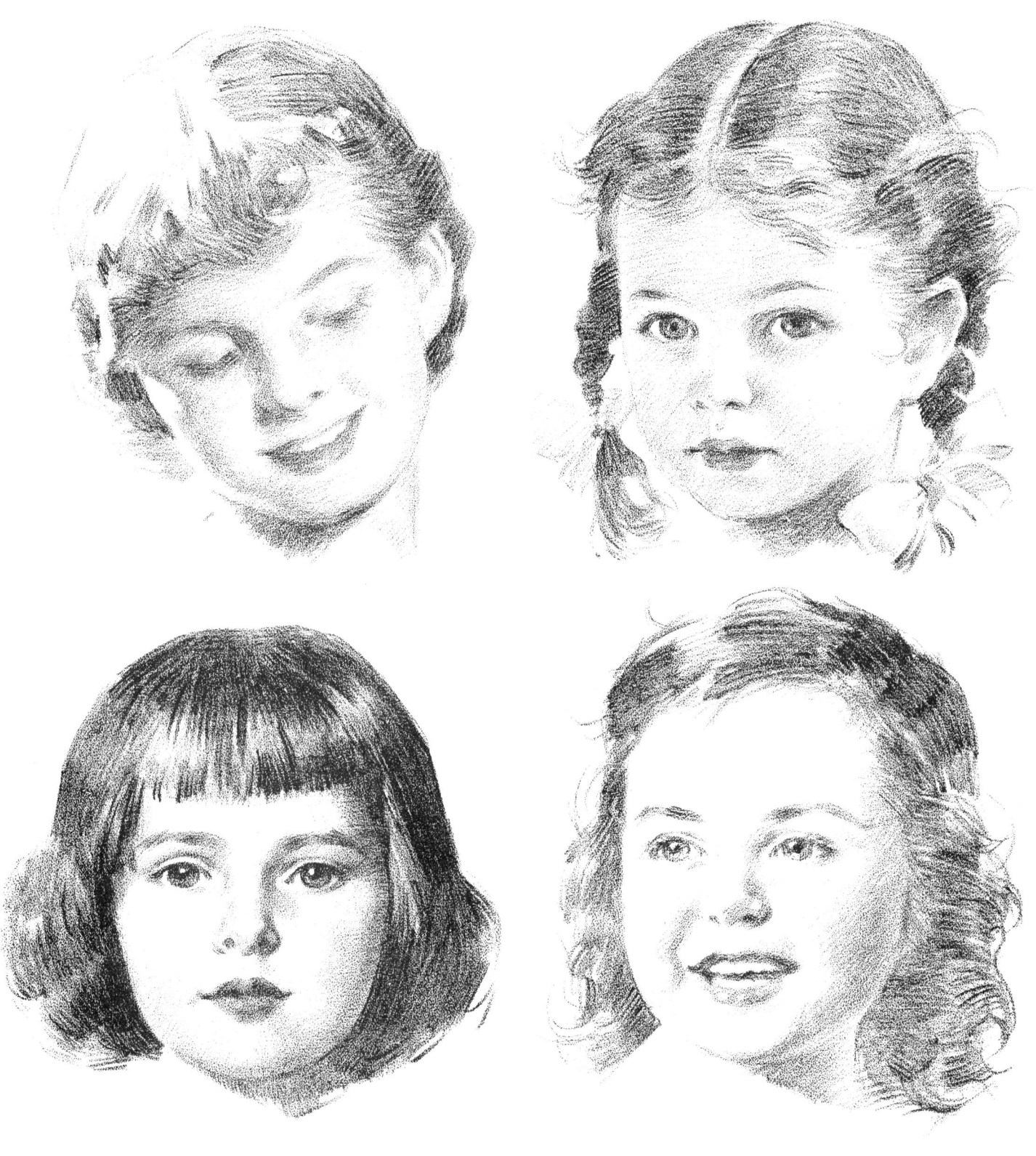

PLATE 64. Studies of little girls

The treatment of the hair has a lot to do with the appeal of a little girl's head. Little pigtails will probably never go out of style. Bangs also seem to be ever popular, and hair hanging loose or in curls is always in evidence. In color drawings or paintings, a bit of color in a hair ribbon is always effective.

109

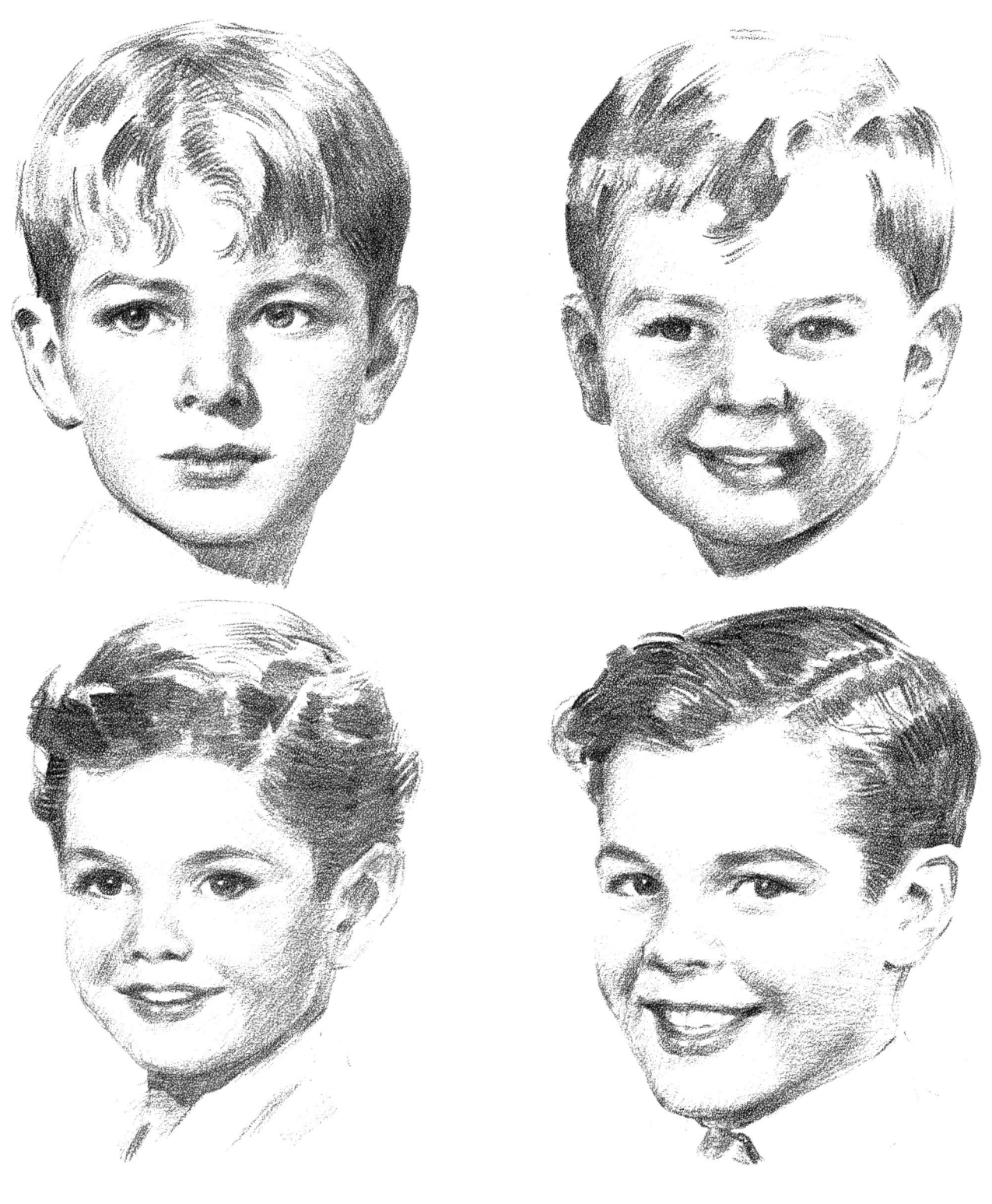

PLATE 65. More little boys

As one progresses in the drawing of children, he becomes impressed with the distinctive character and personalities he finds. Children register as many feelings and emotions as adults, and much more freely and obviously. As we grow older we learn to hide our real emotions, sometimes too deeply. Most children are much more truly themselves than adults are.

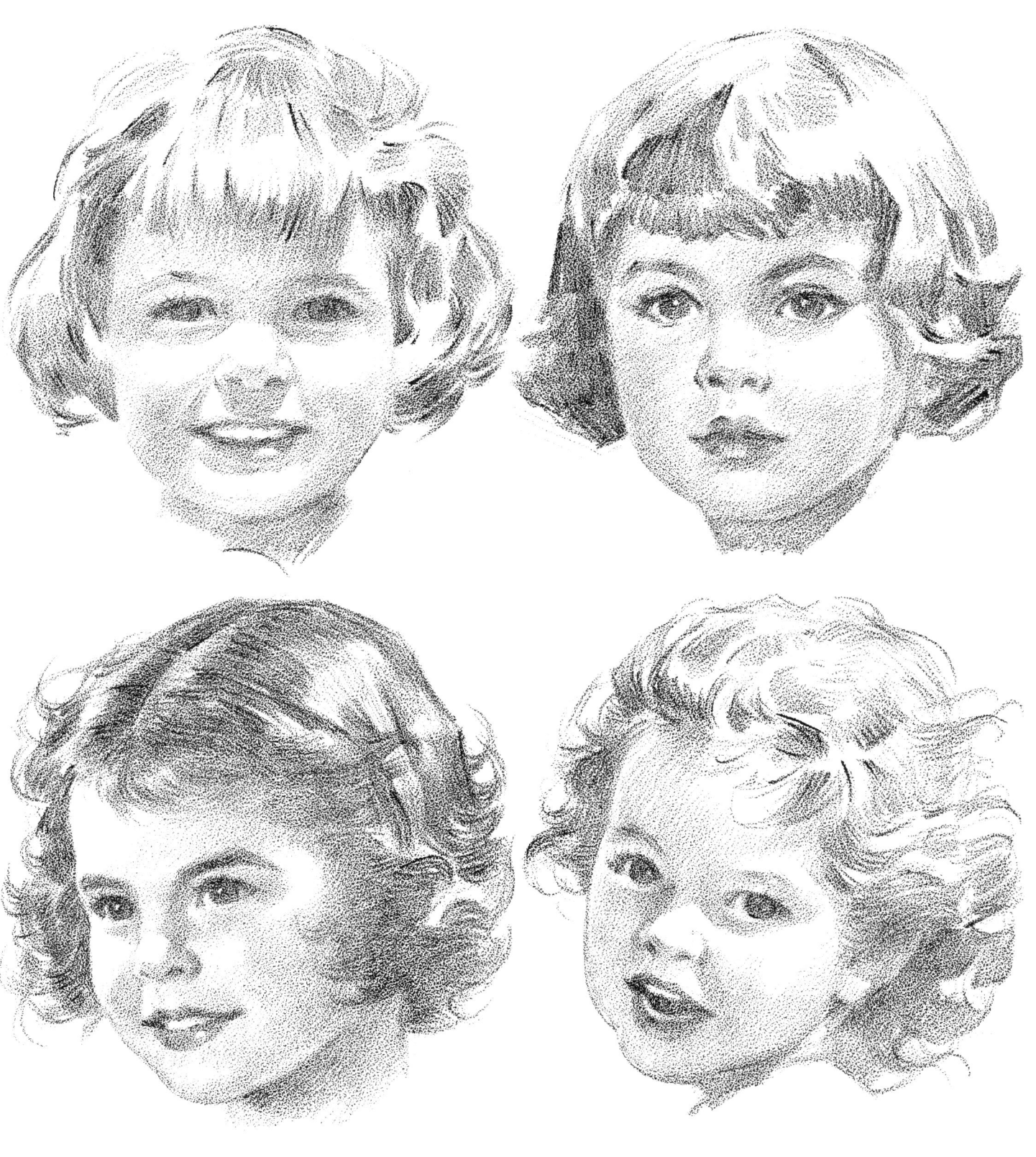

PLATE 66. More little girls

It is much easier to show a child's expression in a drawing if we catch it first with a camera. Their changes of expression are lightning fast, and no child should be asked to hold an expression.

111

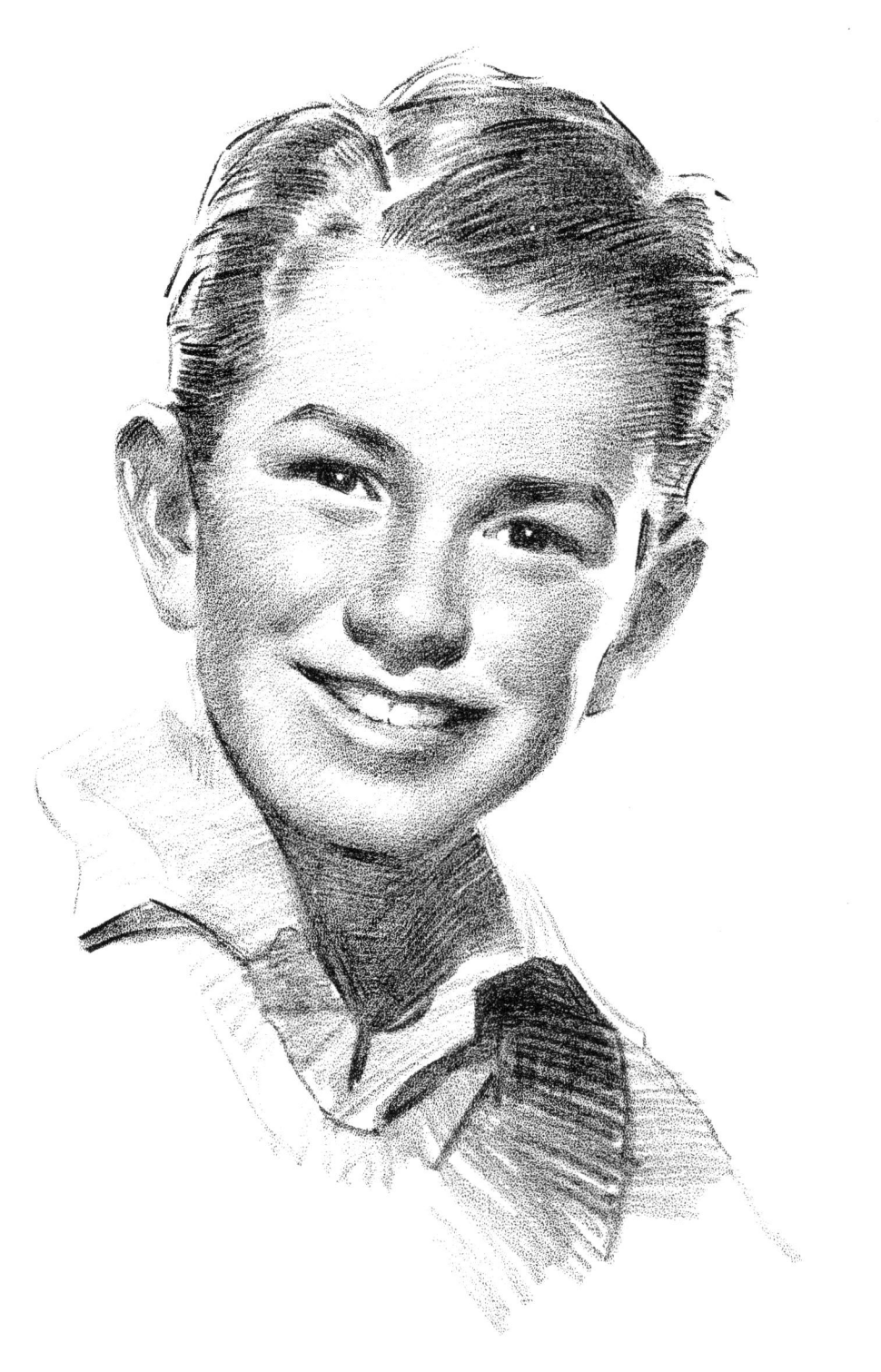

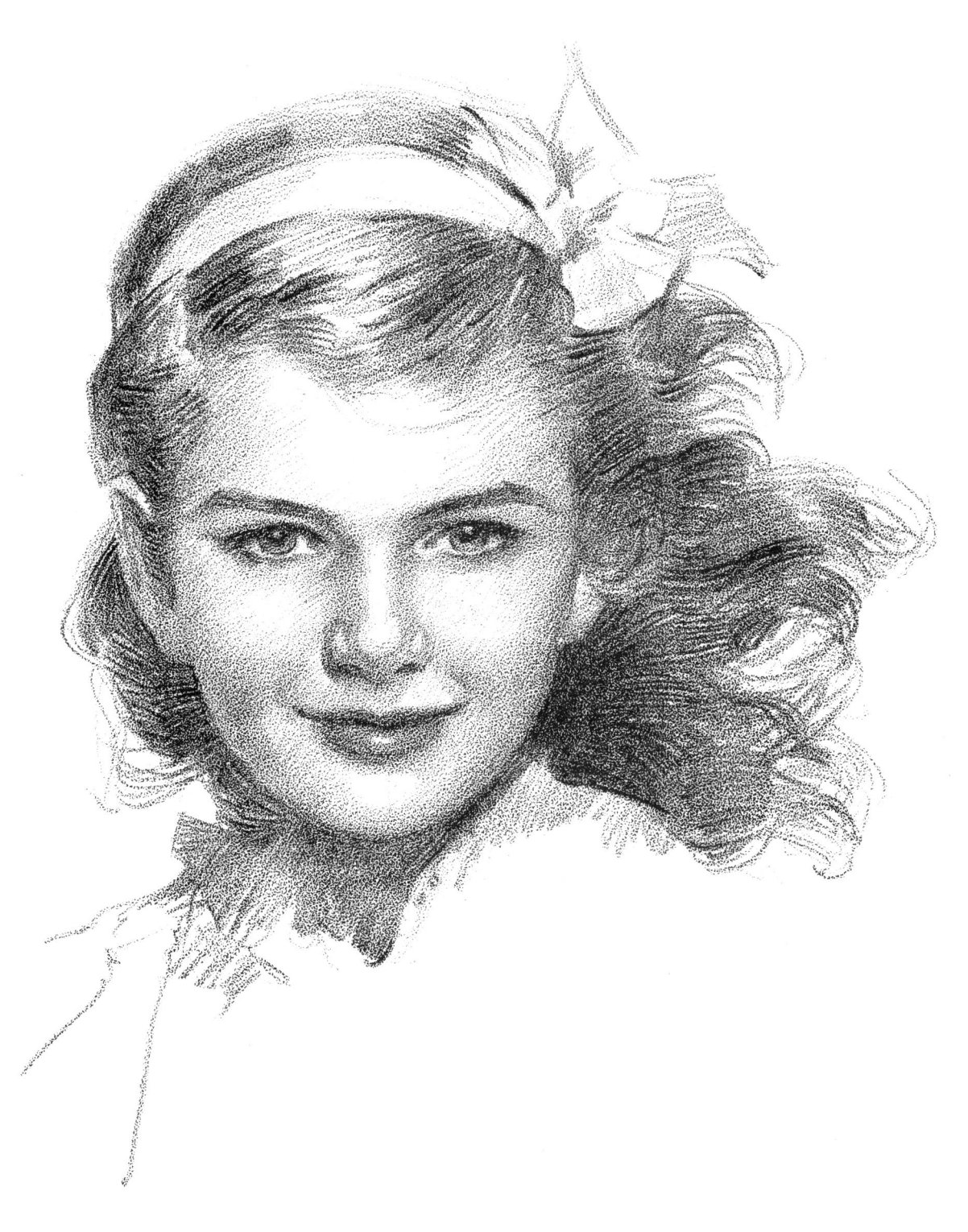

II. SCHOOL CHILDREN

II. SCHOOL CHILDREN

This section deals with children of the grammar-school age, or up to adolescence. That is the age of activity and rather gradual growth, before the spurt of growth that comes at the time of adolescence. It is also the age in which habit and character begin to be formed and to show in the face. We might also call it the age of mischief, because the energy cannot be confined to growth and overflows into physical activity.

It is most important to learn to draw children of this age with a smile—not only on the face you are drawing, but on your own face. Almost one hundred per cent of children in advertising must appear as both active and happy. On the other hand, a youngster's face can be particularly beautiful in repose. Sometimes you will wish that the editors and art directors appreciated this more often. At least when a story is touching, the child may be drawn without a grin. But in advertising, especially of foods, children have to be shown going into ecstasies over the product.

Children at this age live in a world of their own. Most of the time a little revolution seems to be going on inside them, against all the authority which is heaped upon them by parents and teachers and which they are not quite old enough to understand. Try to remember your own schooldays. When asked why you did this or that, you could hardly have answered, "Because I'm getting tired of so much authority." Sometimes adults find it hard to understand why

the effect of our authority slips off so easily, and the answer can only be that there is so much of it.

While we consider this the age of learning, we are likely to forget that much learning is gained by experiment, and not all by direction. All the wonders of invention are holding themselves out for inspection by the young. If your boy takes your alarm clock apart, or strews your pet tools out by the back fence, this comes under the head of experiment without direction, and you would have a dull boy if he didn't do a few of these things.

When drawing children, or even when photographing them, forget that you are grown up. Try hard to meet them in their own world, and draw them out. A child who is afraid of you or who shuts you out is not going to be himself, and so will not be a good model, if you are interested in conveying the spirit of childhood. That spirit lies in their faces only when they are free of authority. Watch their faces change when authority descends on them. I am not speaking against authority itself; I just mean that it does not photograph well, and resentment or sulkiness certainly does not make an attractive picture.

Since proportions have already been thoroughly discussed, you can learn from Plates 67 and 68 to apply them to the faces of school children. It is helpful to understand them, but merely to get them right is not the ultimate objective.

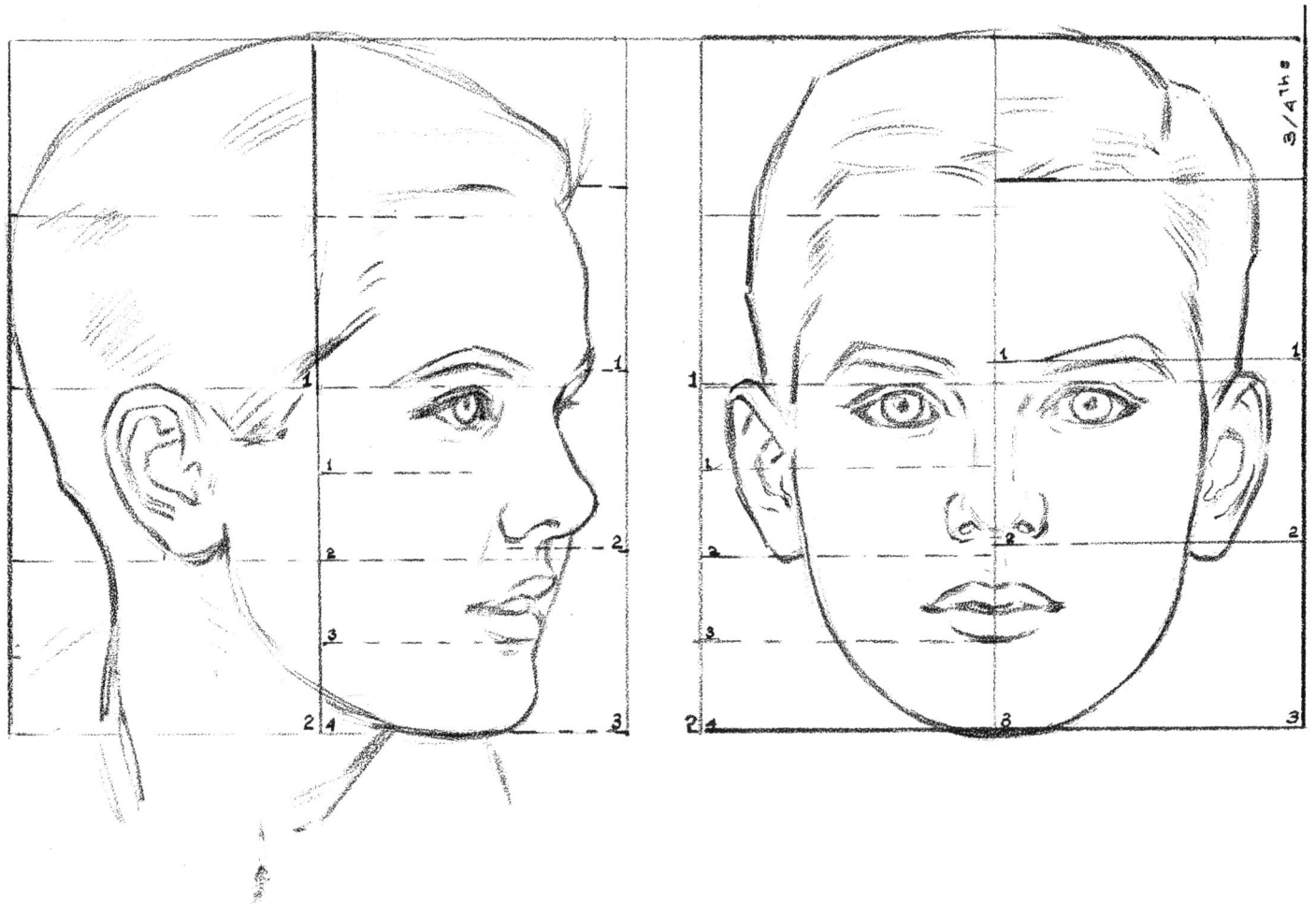

PLATE 67. Proportions of the schoolboy's head

Children between eight and twelve are more difficult to draw than either very young children or adults. The character of the head is pretty well established by this time, and some children have even taken on quite an adult look. But there is a trick to indicating this age group which is quite dependable. The eyes have moved up to touch the halfway line, and the space from the hairline to the top of the head is three-fourths of a unit instead of one-half unit as it is in the adult. In the adult the halfway line cuts through the middle of the eyes and out through the outer corners, while in the child approaching teen age the whole eye is below this line. The nose is still slightly above the second quarter division in the lower half of the face. The lower lip touches the line of the third quarter division.

In boys there is notable development in the ears. The mouth loses much of the baby look. The second teeth have replaced the baby teeth and the jaw has developed to accommodate them. The nostrils develop and the cartilages of the nose spread. The bone at the bridge of the nose develops a little more slowly, so many boys retain a turned-up nose until they are well into their teens.

This is the age of freckles. It is also the age of mischief and carefree happiness, as the expressions show. The hair is unruly; the front teeth look large. While the front of the jaw develops, the rear of the jaw at the corner below the ear does not develop until later. A large square jaw does more than any other feature to give a look of maturity. If you want to keep the face young, keep the corners of the jaw rounded.

116

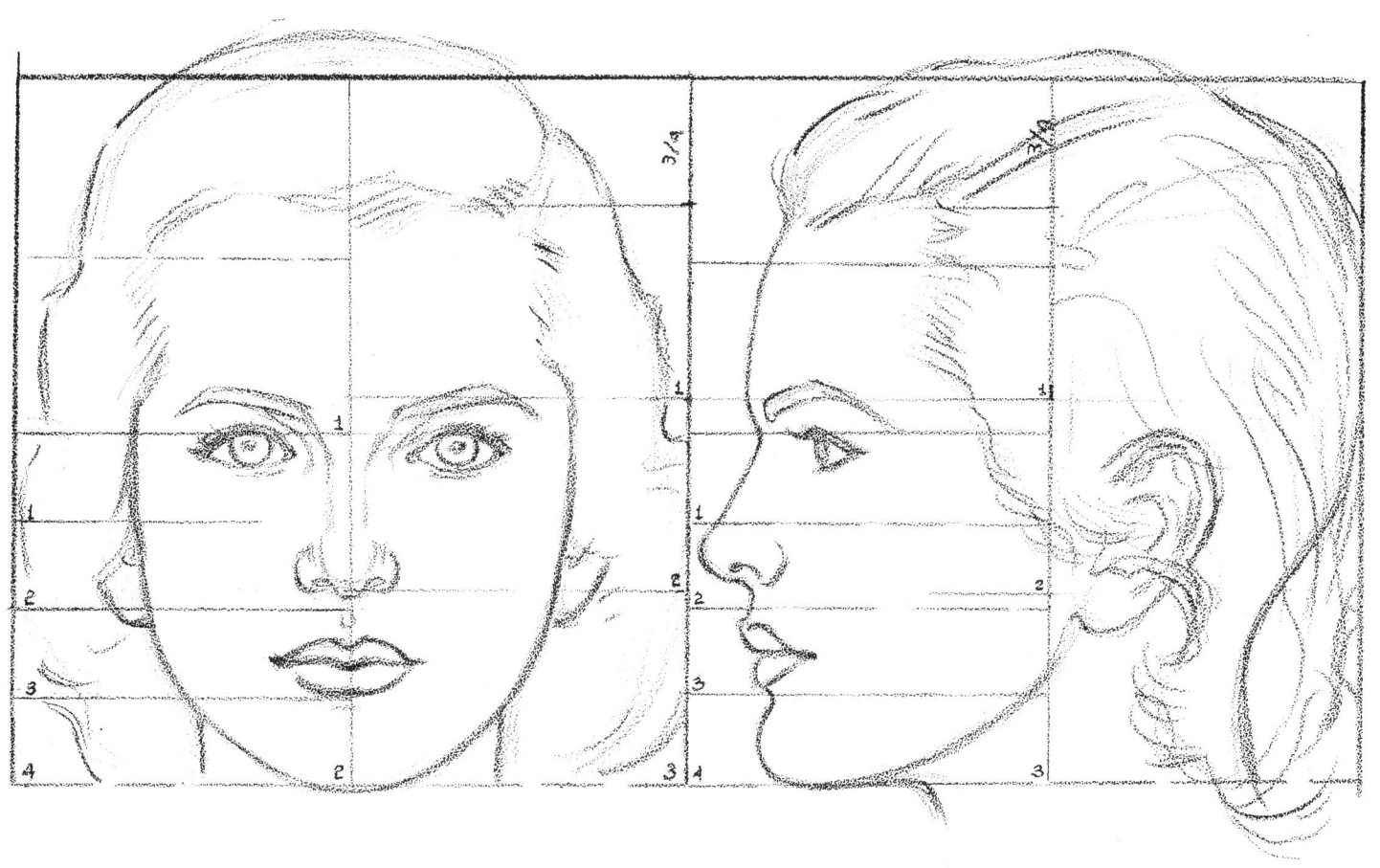

PLATE 68. Proportions of the schoolgirl's head

Young girls seem to mature faster than boys as far as facial characteristics are concerned. Most girls acquire a fairly mature look quite early in their teens. As I mentioned earlier, they usually have higher foreheads, and the hairline is well up. The cheeks are rounder and there is often more space in the front view between the corners of the eyes and the edges of the face where the ears attach.

It must be remembered that here we are dealing with averages. There are always variations and exceptions. Photographs of girls ten to twelve years old often look more mature than the children actually look. Sometimes this is because we are seeing only the head and shoulders, and not the head in association with the rest of the body. In a girl of thirteen or fourteen the head is almost full grown, while the body is not.

Full lips are always appealing in the face of a young girl, and roundness rather than boniness. Girls as well as boys often have freckles at this age, but do not overdo the freckles in drawing girls.

To draw heads of children of this age group well, you will have to practice on a great many.

117

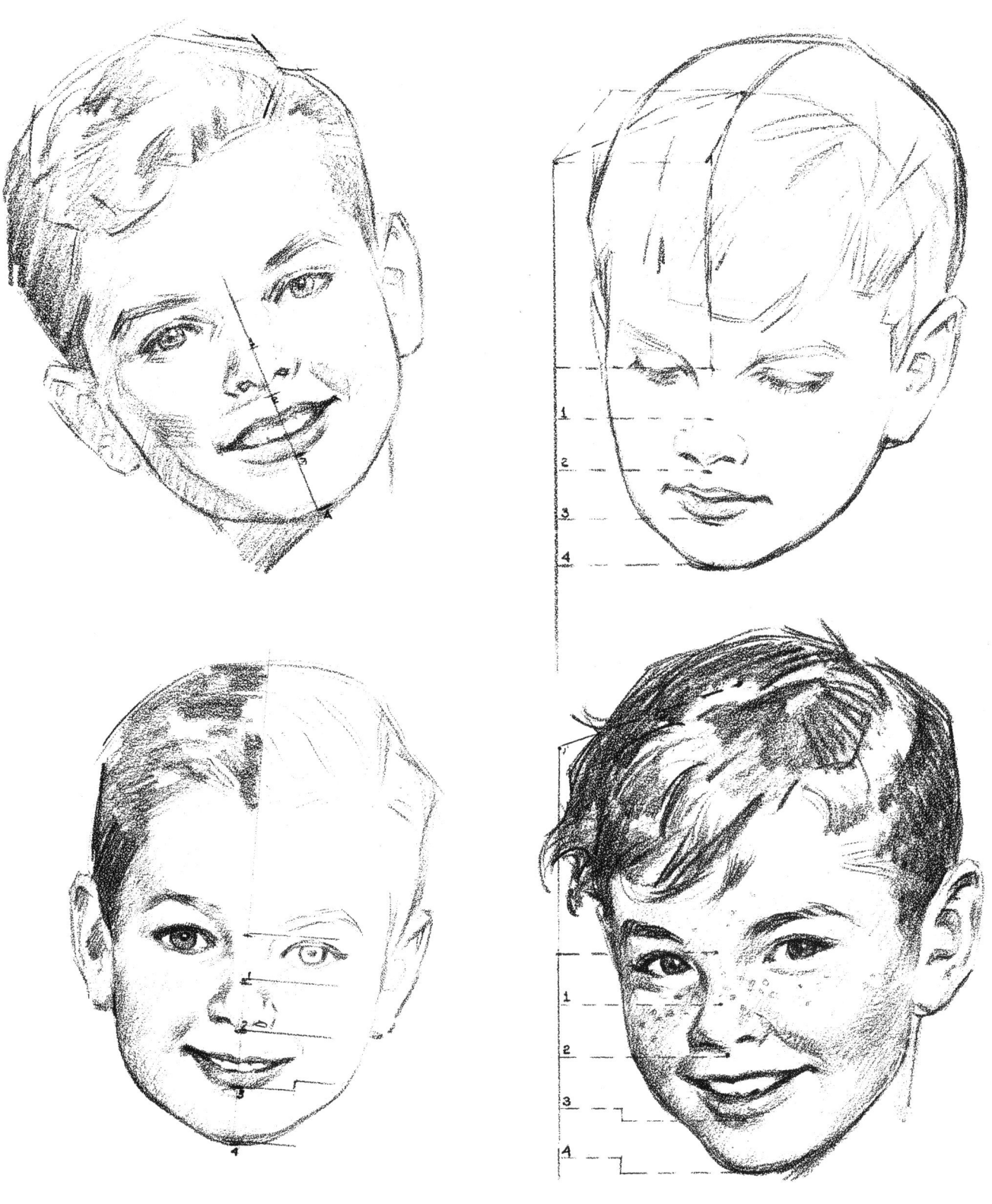

PLATE 69. The four divisions—schoolboys

If you plan to do advertising illustration, or are already in that field, you will find drawing growing boys and girls very remunerative. Practically all foods are advertised to mothers with growing children and the children appear in profusion in such advertising. You can practice from the heads here, or find others in the women's magazines that offer excellent practice.

118

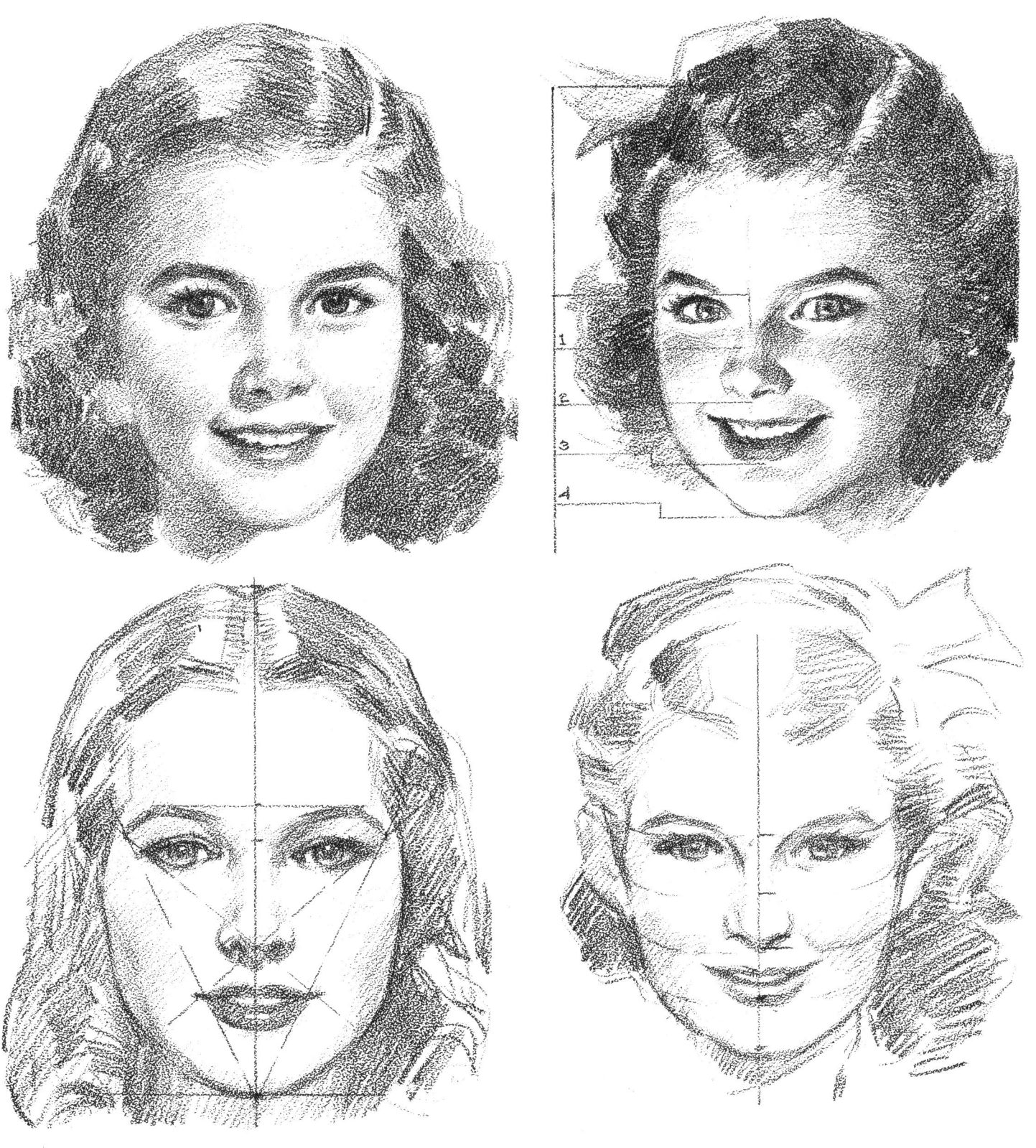

PLATE 70. The four divisions—schoolgirls

At the right, above, we have the usual quarter spacing. It is interesting and helpful to note how the diagonals cross in a young girl's head. The diagonals from the corners of the eyes through a point at the middle of the base of the nose also cut through the corners of the mouth; those from the outer ends of the brows cut through the corners of the mouth to a point at the base of the middle of the chin.

119

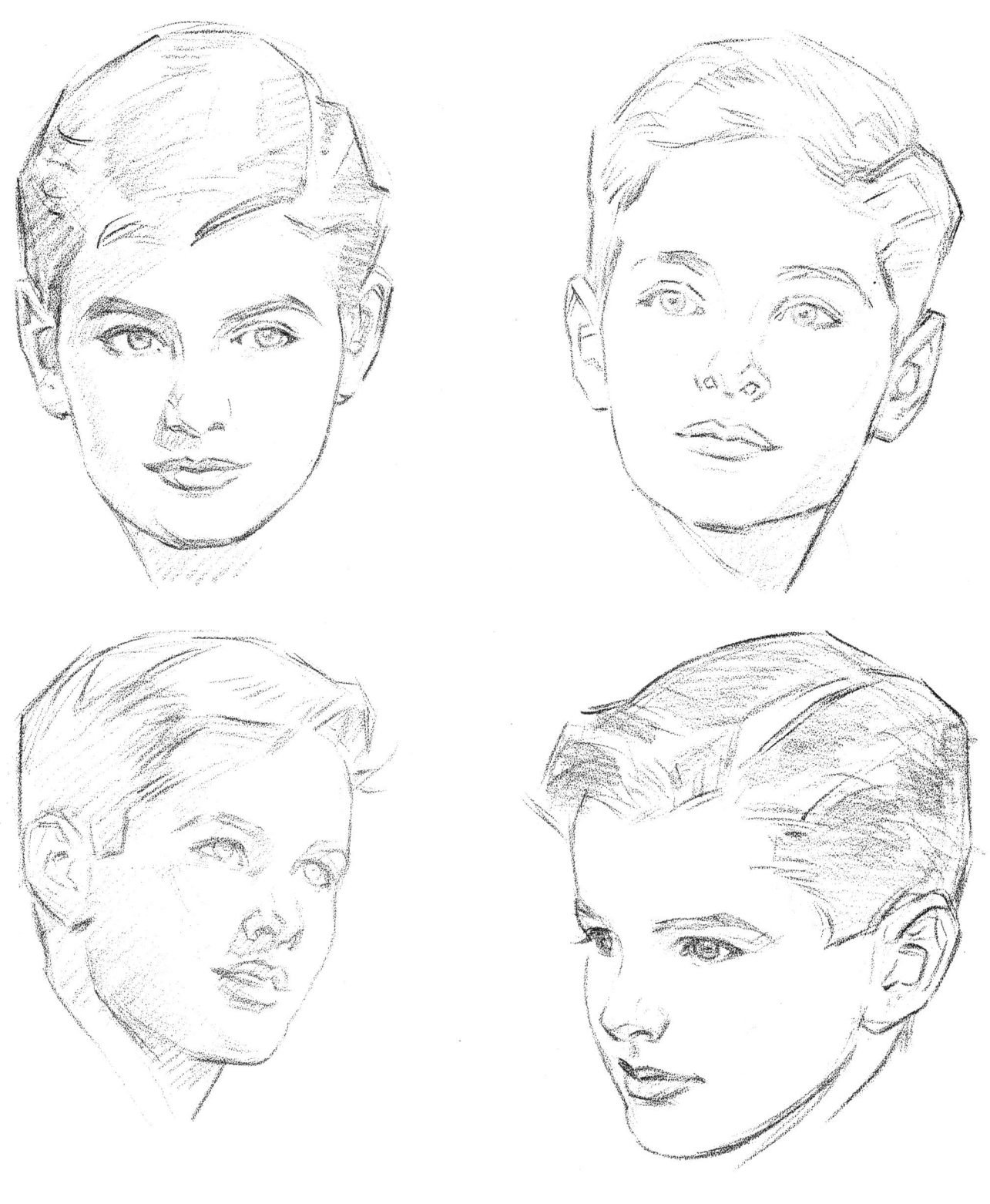

PLATE 71. Sketches of schoolboys

These heads have been left in outline since the outlines will probably be more helpful than the finished heads. There is a wideness to young faces that is more felt than measured. In drawing young people it is particularly important to trust your feelings. Once in a while a face will look older or younger than you intended no matter what you do. In that case the best thing to do is to try another subject.

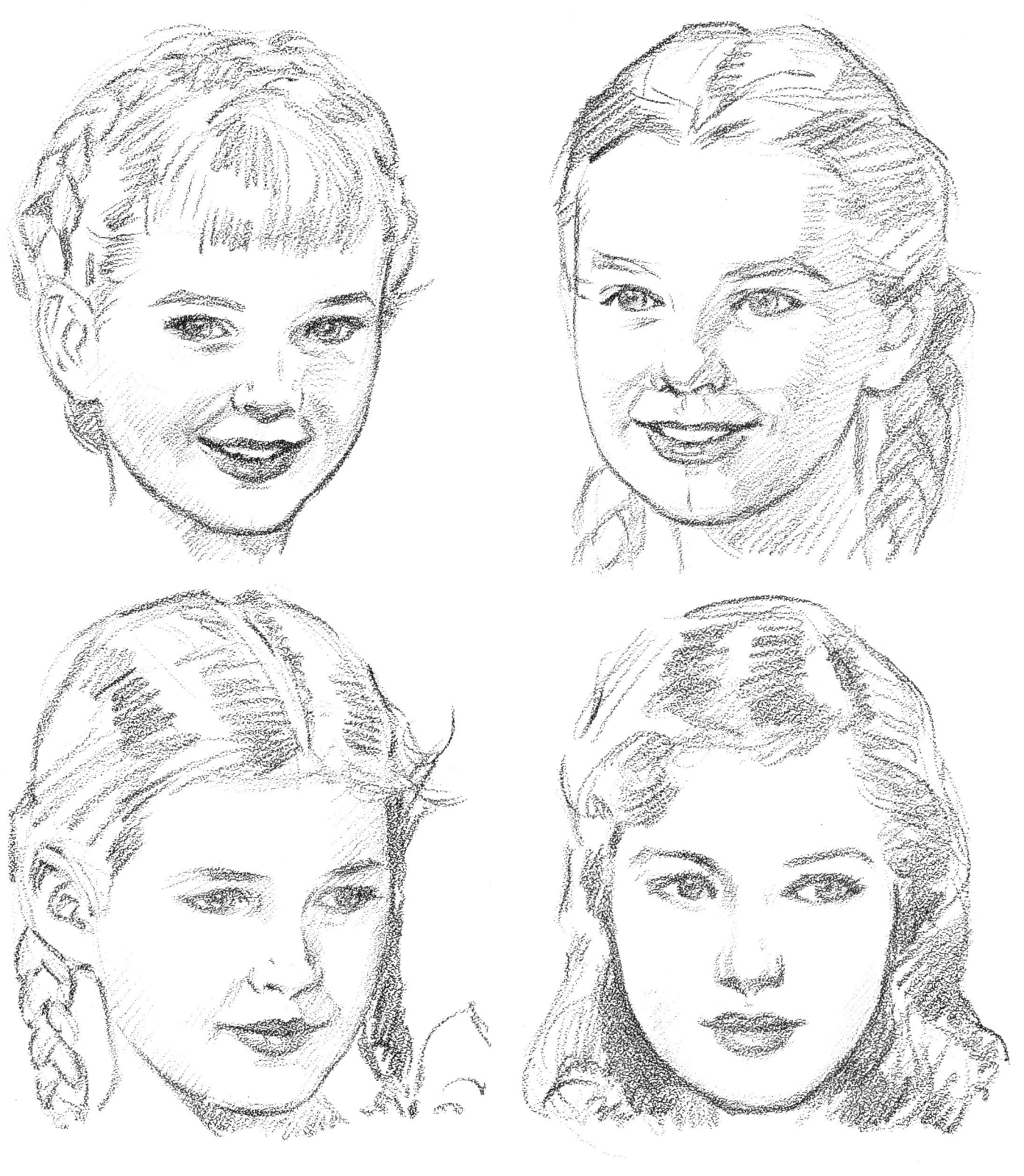

PLATE 72. Sketches of schoolgirls

Draw heads in outline until you are satisfied that the age and expression look right. There is no point in adding tone to a head that does not appeal to you. The tone can only build up the forms already established. If they are wrong, tone does little to help. Sometimes a head in outline may look better than one completely finished.

121

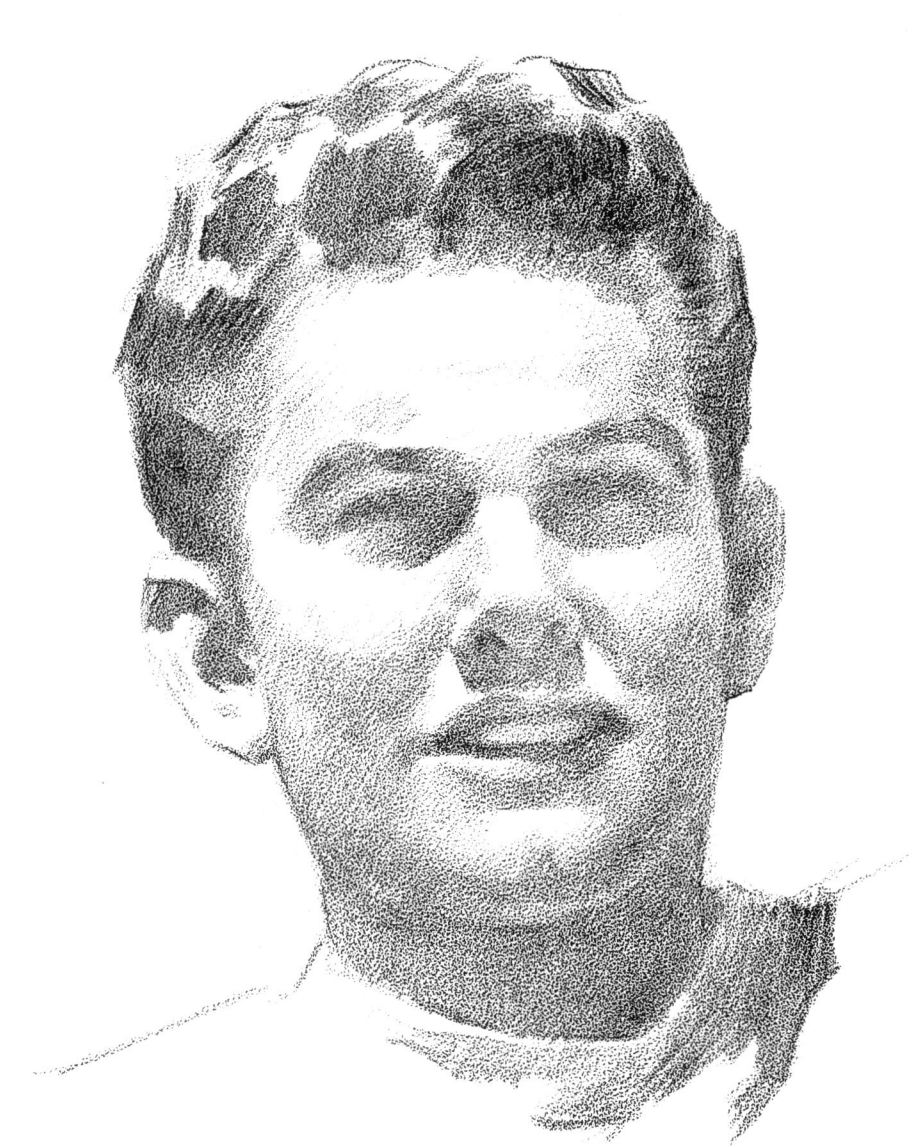

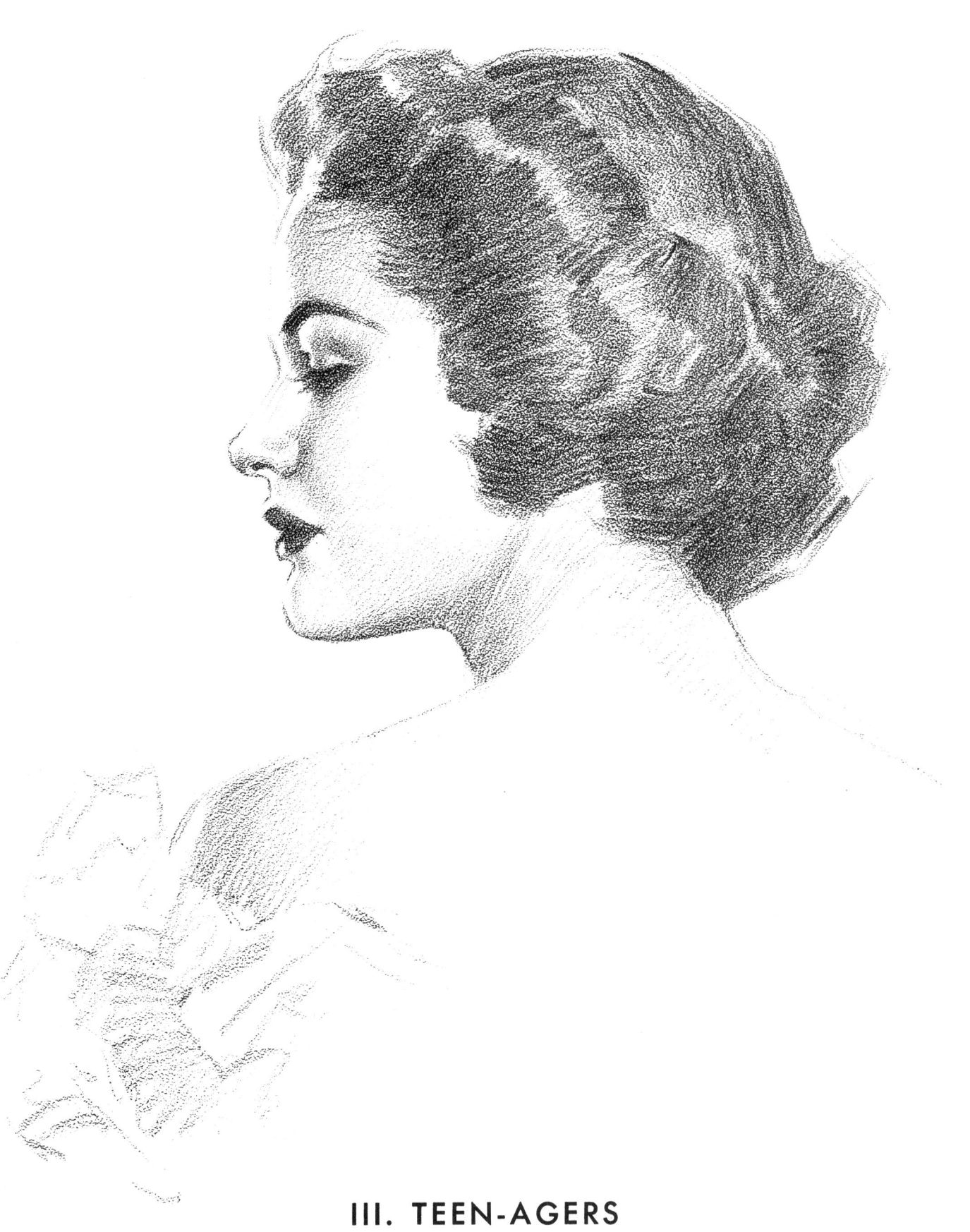

III. TEEN-AGERS

III. TEEN-AGERS

Teen-agers are popular subjects in fiction, advertising, and portraits. Since the proportions of the head are so nearly those of the adult head, we are almost back to where we started, but I hope with much more understanding.

In drawing teen-age boys and girls we must take into consideration the great variety of types. In boys, bony faces with well-marked muscles are associated with athletic types. The muscular activities contribute to a certain leanness. Some boys grow so fast they are robbed of some vitality; others simply do not lean toward athletics. Another type of teen-age boy has a round face, long legs and arms and large hands and feet, tends to drape himself over anything suitable to rest upon, and hates effort—especially home chores. As a rule, these boys develop more energy later when they attain full growth.

Since most teen-agers—girls as well as boys— are big eaters, if they do not exercise, they have a tendency toward fatness. Fortunately, they lose most of this excess weight in the spurt of energy that follows full growth.

Treat teen-agers with as much understanding as possible. Remember that this is the age of the first big heart throb, the age when the urge to be different from their elders comes out in every conceivable fad, in dress, hair-do, and personality. Study teen-agers closely to catch the spirit, for youth is elusive in more ways than one.

Now that we are completing our study of heads, you will find it rewarding to review parts of this book which might have given you trouble earlier. The new drawings should show great improvement over your first ones. You will find everything much easier, and will also have gained confidence from your practice work.

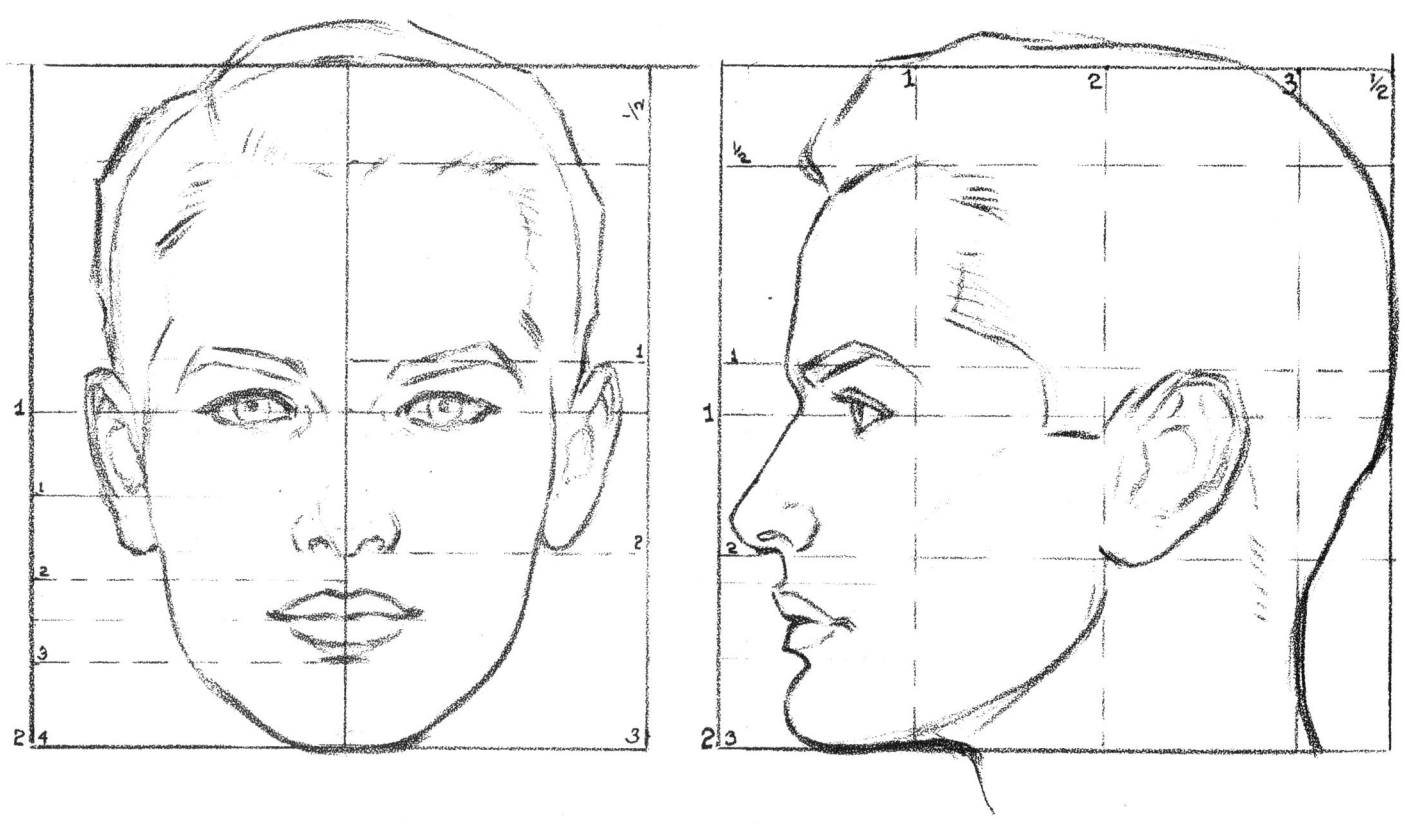

PLATE 73. Proportions of the teen-age boy's head

The proportions of the head in teen-agers are almost identical with adults; the difference is largely a matter of feeling. In boys the bone structure has become quite evident, though it should not be stressed as much as in men's heads. There are no noticeable lines. The flesh is firm and still inclined to smoothness. The cheeks are smooth without much definition of the muscles. The jaw has developed considerably in a short time. The bridge of the nose has taken permanent shape. As the jaw and cranium have grown, the ears appear smaller in relation to the whole head than they do in a little boy. The cartilage of the ear is now well defined; the ears have lost much of their roundness and taken on more angular lines.

The hair has moved back somewhat from the temples. The brows have definitely thickened. The lips are fully developed in size. The chin has come forward in permanent shape.

The only bone not fully developed is the corner of the jaw. This continues to develop, research shows, until the age of twenty or more. I suspect the cranium itself does not reach its maximum growth until full maturity, though further growth does not perceptibly affect the proportions of the head.

126

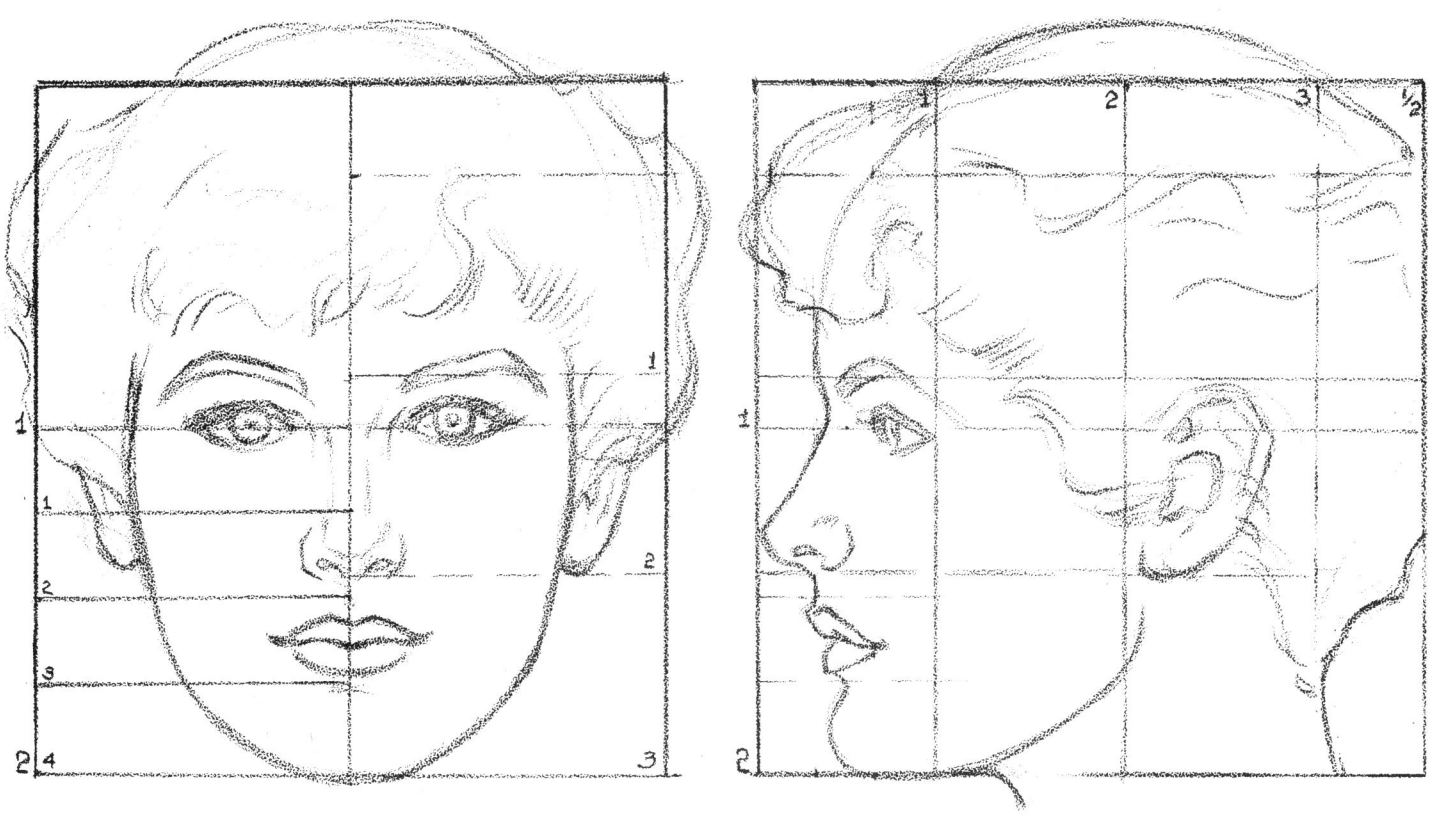

PLATE 74. Proportions of the teen-age girl's head

Sixteen is traditionally the perfect age for girls. By that time they have lost the gangliness of fast growth, and all is smooth, round, and fair. Now that girls also engage in athletics, their faces tend to show more muscle than did those of their mothers at the same age. But the predominating quality is youth—the faces are unlined, full of freshness and vigor.

These things are important in portraying young people, because the actual proportions of the face change very little from sixteen to sixty. The jaw in the girl may develop a little, but hardly enough to affect the drawing of the proportions much. That is why the artist must more or less "feel" the age he wishes to draw.

It is quite important to obtain good material to work from. Faking a drawing of a beautiful young American girl is a very difficult thing to do, until you have drawn a great many heads, and know the basic construction inside and out. I do not believe any of the outstanding artists proceed without adequate material to work from. Beauty, remember, is largely a matter of perfect proportions and perfect placement of features. The commercial illustrator will need to draw many pretty girls.

127

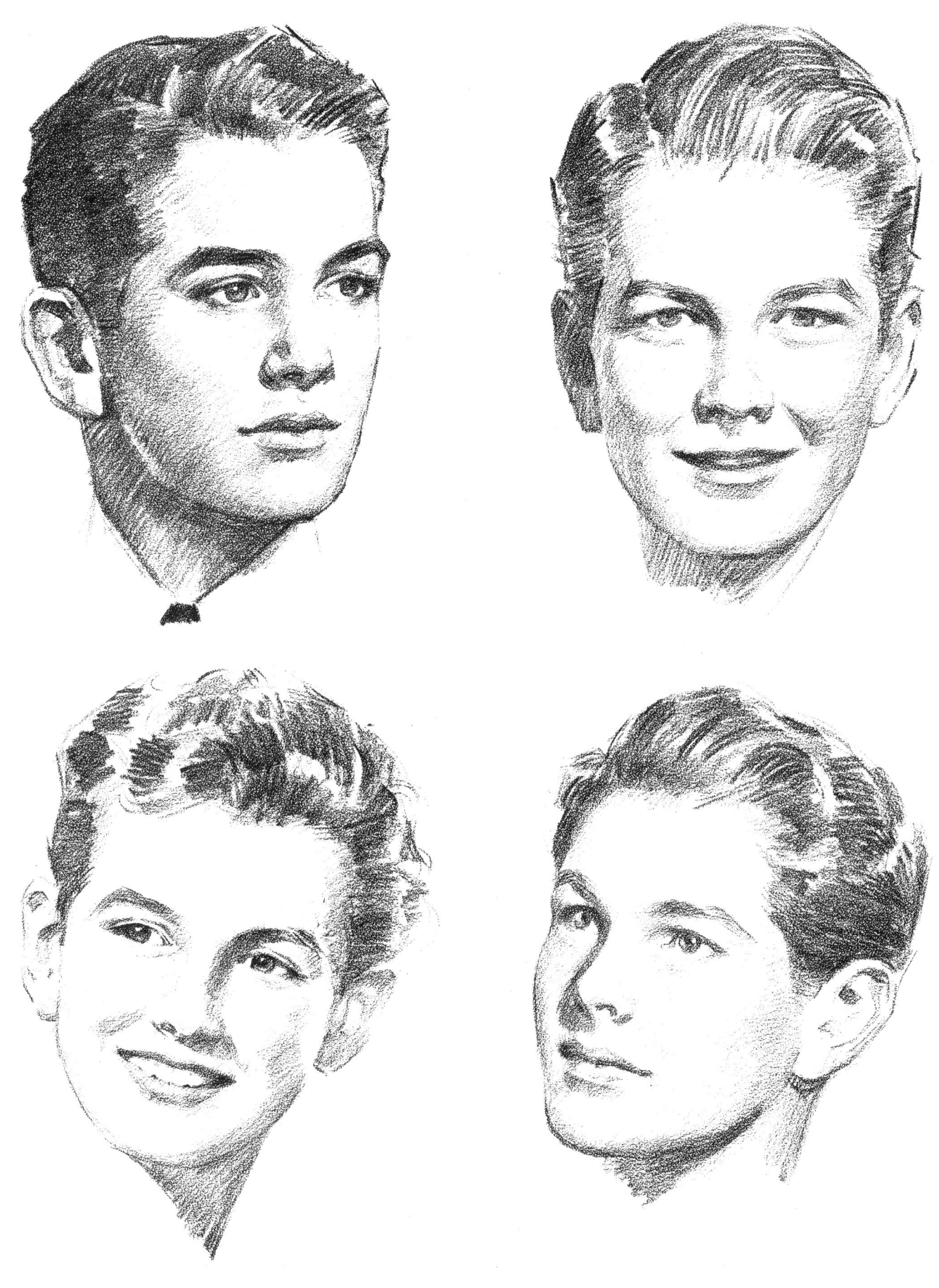

PLATE 75. Teen-age boys

128

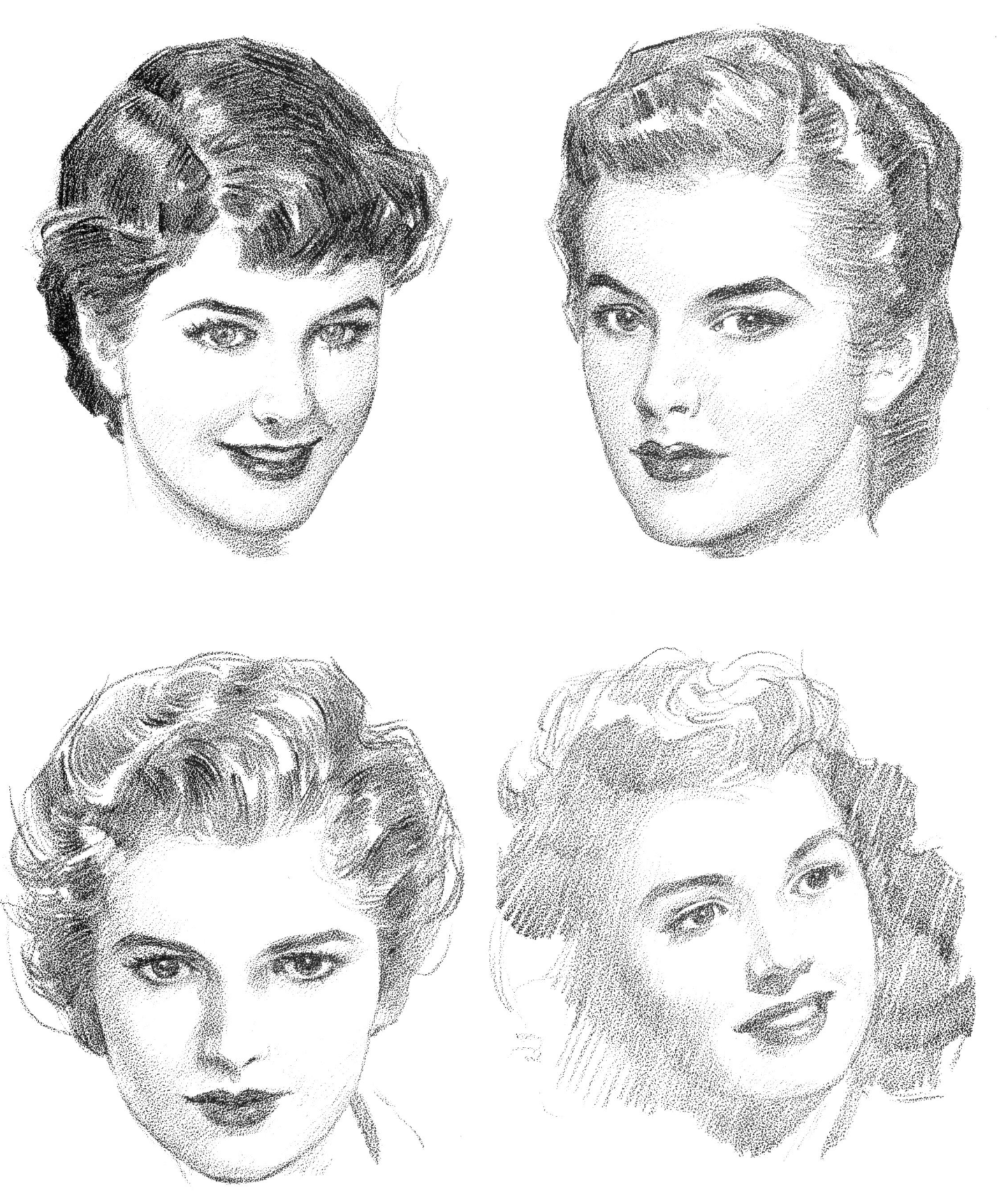

PLATE 76. Teen-age girls

129

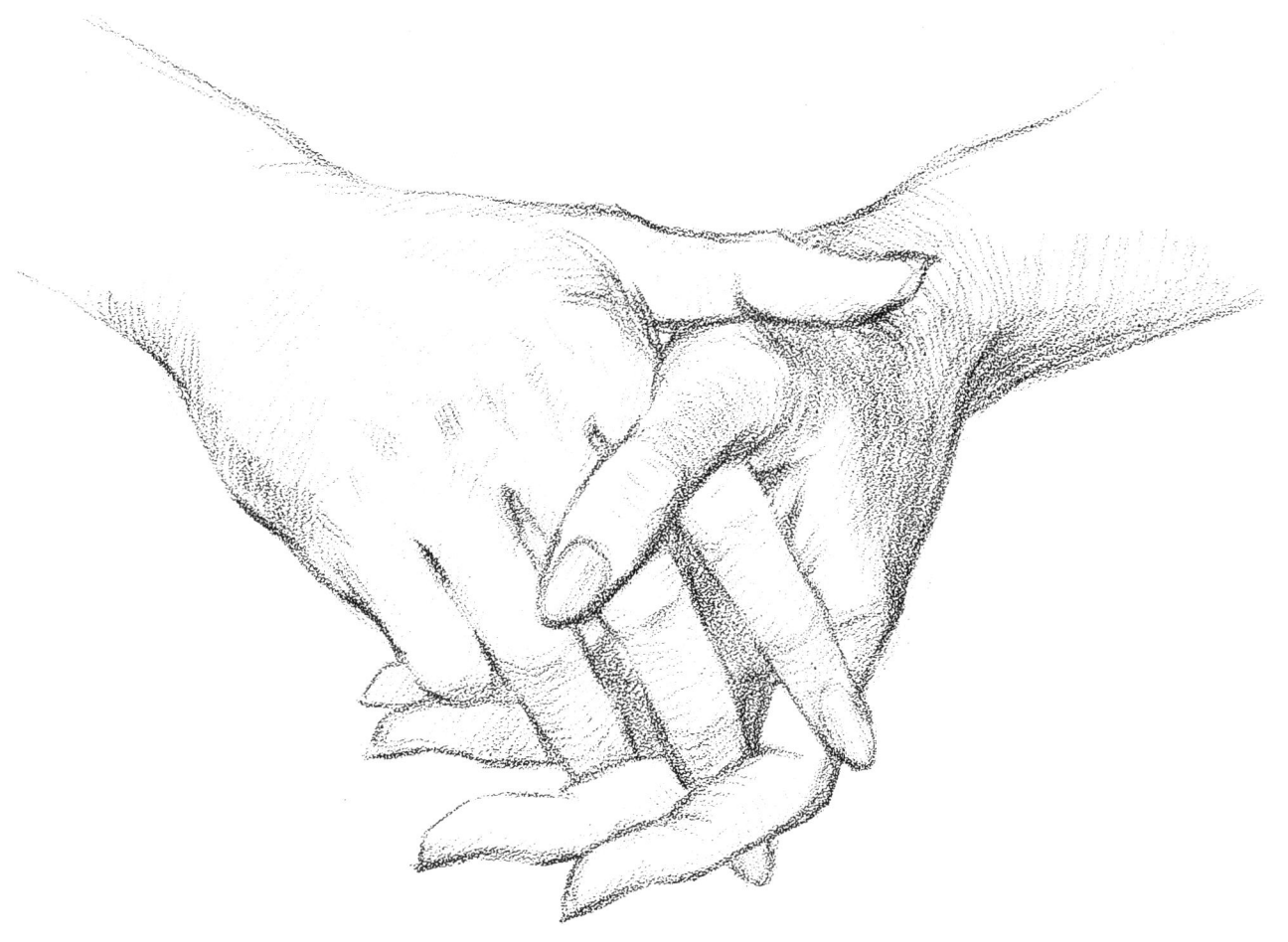

Part Five: Hands

Part Five: Hands

PERHAPS NO ASPECT of drawing is accompanied by more confusion and provided with less adequate material for study than is the drawing of hands. Much of the trouble is caused by searching for material instead of using the material you have available, because in your own two hands you have the best source of information available. Perhaps you have never thought about them in that light. Drawing of hands must be largely self-taught. All any instructor can do is point out the facts that lie right in your own hands.

The study of hands, aside from learning their anatomical construction, consists mainly of breaking down the measurements of various parts into comparisons. Fingers have a certain length in relation to the palm; spaces between the joints of the fingers are in definite proportion to the whole finger. The palm is so wide in comparison to the length. The distances between the knuckles on the back of the fingers are longer than those between the creases on the undersides. The length of the longest finger from its tip to the third knuckle in back is practically half the length of the back of the hand from fingertip to wrist. The thumb reaches nearly to the second joint of the first finger. The length of the hand is about equal to the length of the face from chin to hairline. You can make these comparative measurements as well as anyone else.

The hand is the most pliable and adjustable part of the whole anatomy; it can be made to fit around or grasp almost any shape within reasonable size or weight. This pliability is what causes difficulty for the artist, because the whole hand can assume countless different positions. Yet the mechanical principle by which the hands work remains constant. The palm, as a hollow, opens and closes, and the fingers fold inward toward the middle of the palm. The

nails are really a stiff backing for the tips of the fingers, as well as an extra edge for precise grasping. You pick up a pin with the fingertips; you pick up a hammer with the palm and fingers. The back of the hand is more or less rigid to the backward pressure of the fingers, as used in pushing. For adjustment to almost unlimited purposes, the hand is the most wonderful mechanism we know. In addition to its perfection as an instrument, it is perhaps more closely coordinated with the brain than any other part of the body is. Many of its movements are controlled by subconscious reflexes; examples are typing and playing the piano.

Man started to educate his hands long before he educated his brain in the cultural sense. The infant can use his hands effectively long before he can think. He will grasp a lighted match before he has learned that it will burn. The story of man's progress from prehistoric times must be closely associated with the adaptability of the human hand.

The fact that the hands and their movements require so little conscious thought may be one reason why so little thought is given to drawing them. Look now at your own hands; you will see them in a new light. Note how the hand automatically assumes a shape compatible with an object *before grasping the object*. To draw a hand in the act of picking up an object you must first study the contour of the object, then observe the automatic adjustment of the hand to fit that contour. Start to pick up a ball, a peach, or an apple and watch your fingers adjust themselves, just ahead of the grasp. The mechanical principle involved is very important in the drawing of the hand. Only by knowing how it actually works can the hand be drawn convincingly.

The back of the hand can usually be drawn in three planes—one for the thumb section as far as the bottom knuckle of the first finger, and the other two across the back of the hand, tapering to the wrist. In most actions the back of the hand is curved and the curve is reduced to these three planes. The palm is usually the three blocks surrounding the hollow of the palm—the heel of the hand, the thick base of the thumb, and the padded portion just under the fingers. The knuckles of the fingers and thumb must be aligned to work inward toward the hollow of the palm, or when outstretched to be at right angles to the direction of the column of the finger. We must also be careful to align the nails so that they lie on top of the column with the middle line of the nail extended from the middle line of the column of the finger. Otherwise the nail may slip around the finger without our realizing what is wrong.

Keep studying your own hands to learn about hands in general. The inner muscles are so deeply embedded that they are not as important as the outer shapes. The only indication of bone we see is across the back, the knuckles, and the wrists. If you get the shape of the palm in almost any action, the fingers can quite easily be attached to it and aligned with it. Study the comparative lengths of the fingers; remember that the thumb works mostly at a right angle to the fingers. Get rid of the idea that hands are hard to draw. They are simply confusing to draw unless you know how they operate. Once understood, hands become fascinating.

The most important fact to remember about the hand is that it is hollow on the palm side and convex on top. The pads are so arranged around the palm that even liquid can be held in the hand. The hand served primitive man as a cup, and by cupping the two hands together he could eat food which he could not hold with his fingers alone. The big muscle of the thumb is by far the most important one in the hand. That muscle, combined with or in opposition to the pull of the fingers, gave man a grasp powerful enough to hold even his own weight in suspension. This powerful muscle held his club, his bow, his spear. Animals depend upon the jaw muscles for existence, but we might say that man depended upon his hands.

When you have mastered the construction and proportions of the hand (Plates 77 to 85), you will find it easy to use your knowledge to show the special characteristics of women's hands and those of babies, children, and older people.

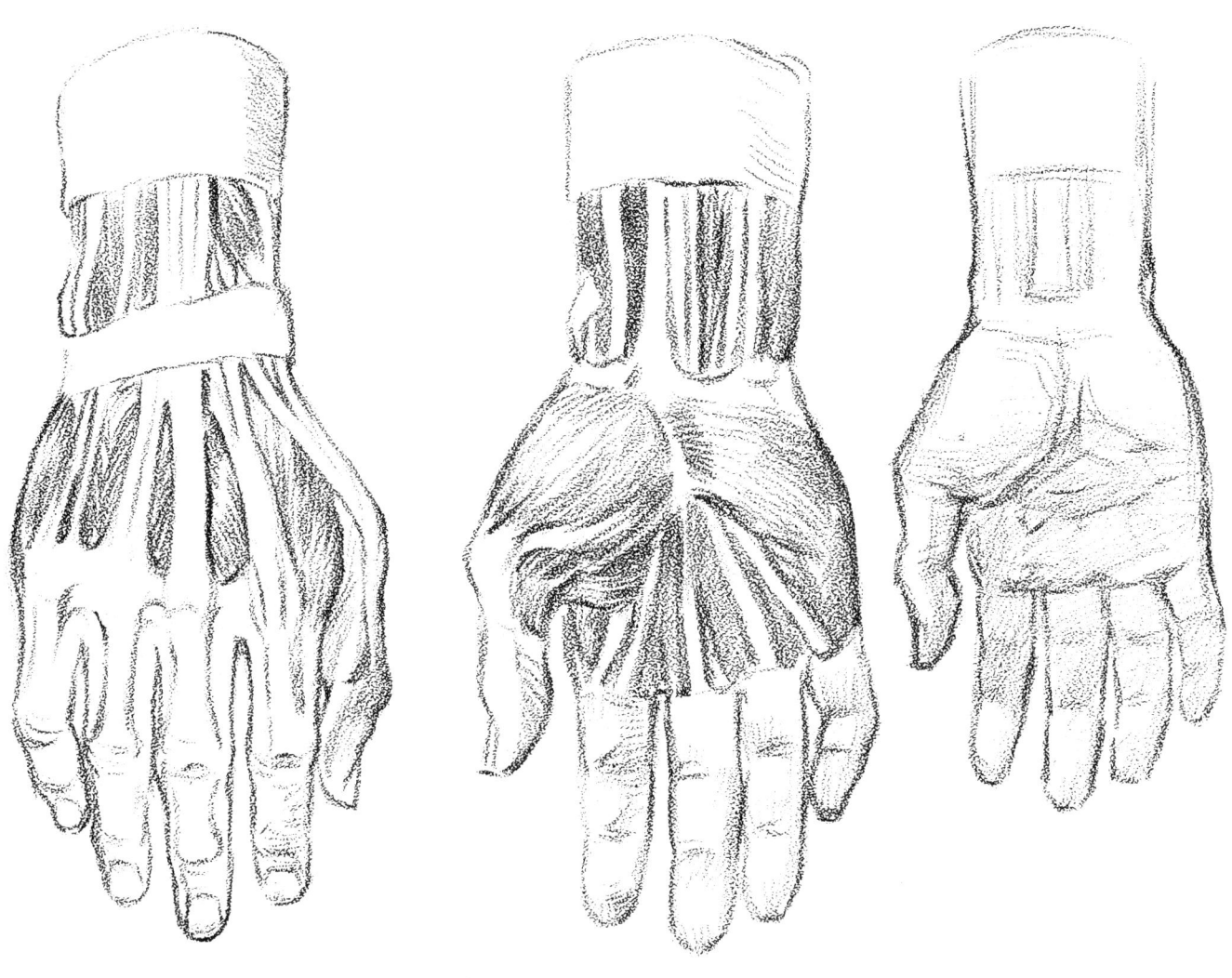

PLATE 77. Anatomy of the hand

Note the strong tendon which attaches to the heel of the hand, and how, on the back of the hand, the tendons are grouped to pull the fingers out. The operation of these tendons is marvelous, for they can operate all the fingers together from inside or outside the palm, yet can control each finger separately. The muscles which pull these tendons are located in the forearm. Fortunately for the artist, most of the tendons of the palm are buried deeply and do not show. In babies and young people, the tendons on the back of the hand are hidden, but they are much in evidence in the hands of adults and the aged.

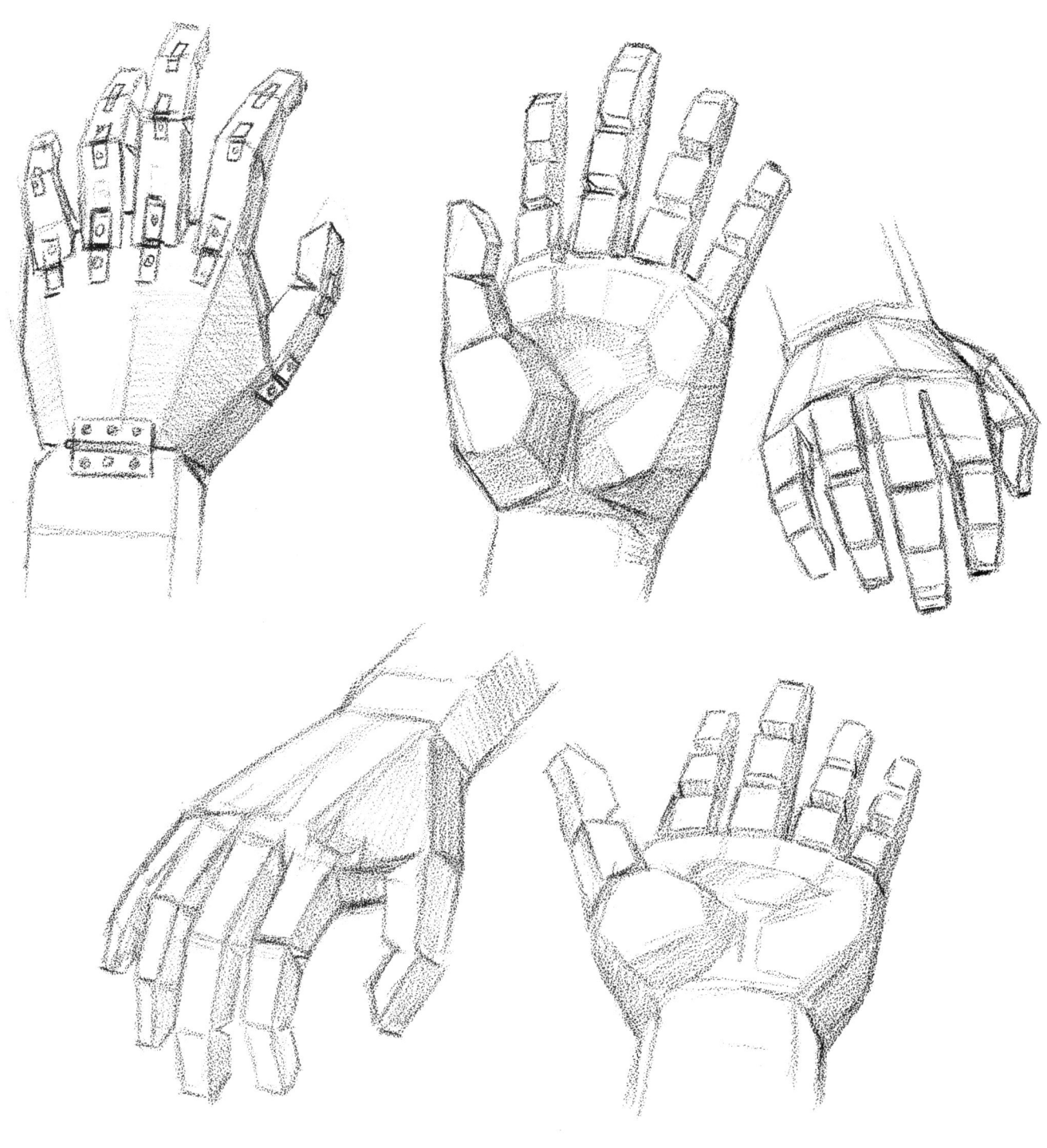

PLATE 78. Block forms of the hand

The bones and tendons across the back of the hand are close to the surface; those around the palm and inside of the fingers are thoroughly padded. I have blocked out these pads so you can familiarize yourself with them. Note the extra thickness of the pads of the thumb muscle and the heel of the palm. At the base of each finger there is a pad. These combine to make a pad across the top of the palm. The pads of the fingers protect the bones inside. Since these pads are all pliable, they provide an even firmer grip on objects much as the pliable treads on an automobile tire grip the surface of a road. There are no pads on the top of the hand, though the pad at the outer edge on the little-finger side can take a tremendous blow, especially with the fist closed, without injury to the hand.

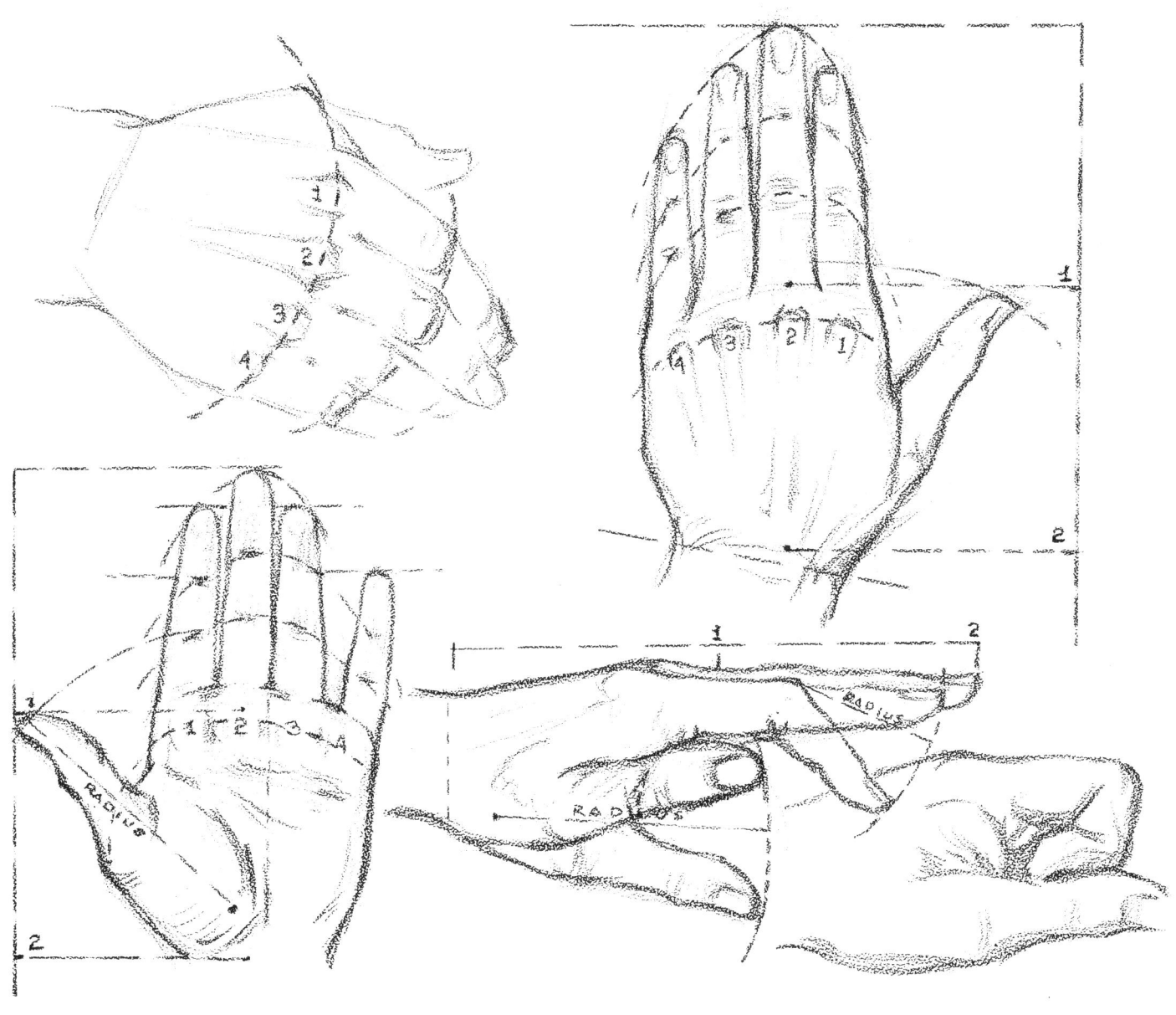

PLATE 79. Proportions of the hand

The next thing of importance is the curved arrangement of the fingertips and knuckles. Two fingers lie on each side of a line drawn through the middle of the palm. The tendon of the middle finger just about divides the back of the hand in half. Important also is the fact that the thumb is turned at right angles to the other fingers. The thumb operates mostly in and out from the palm, while the fingers open and close toward the palm. The knuckles of the fingers are slightly above their creases on the inside of the fingers. Note the flat curve of the knuckles across the back of the hand, with the curves getting deeper as they cross the knuckles toward the fingertips.

The middle finger is the key finger from which we determine the length of the hand. The length of this finger to its knuckle in back is slightly over half the length of the hand. The width of the palm is slightly more than that of half the hand on the inside. The first or index finger just about reaches the fingernail of the middle finger. The third finger is about equal to the index finger in length. The little finger just reaches the top knuckle of the third finger.

137

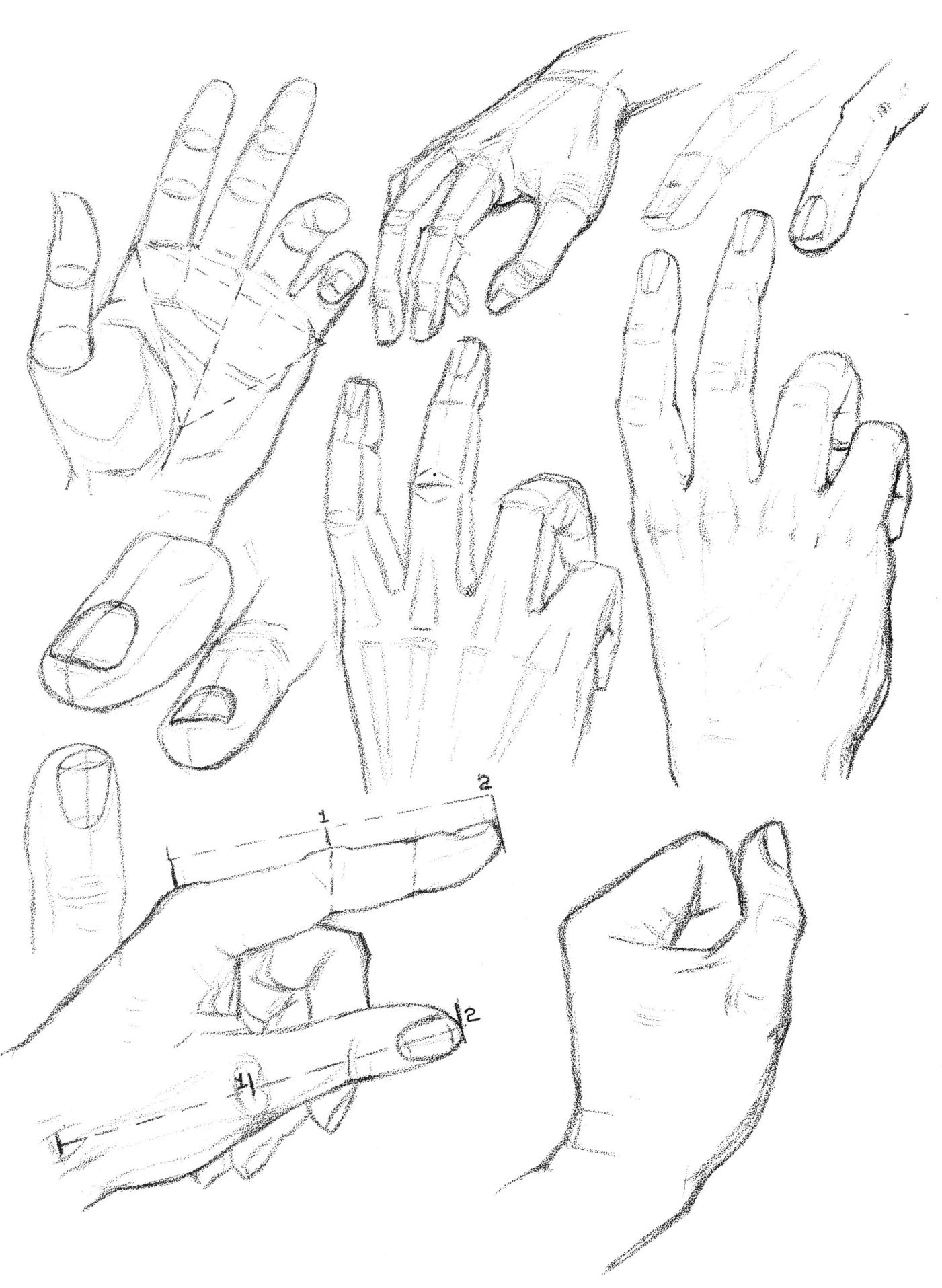

PLATE 80. Construction of the hand

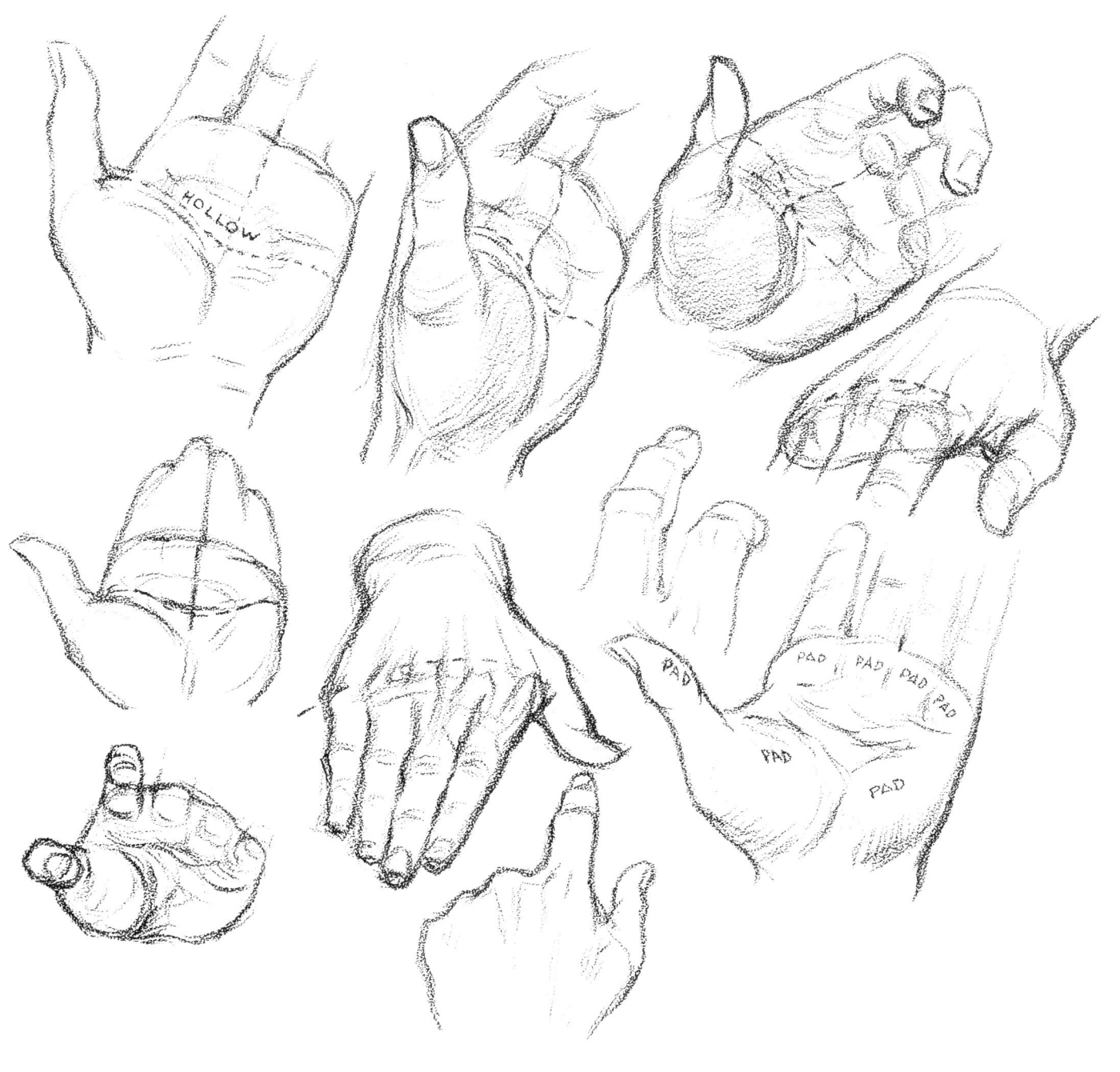

PLATE 81. The hollow of the palm

In the drawings above, note how the hollow of the hand has been carefully defined. Also note the resulting curve of the back of the hand. Hands never look natural or capable of grasping until the artist understands this feature of the hand. All these hands look as if they could take hold of an object. The loud sound of clapping comes from the sudden compression of air between these two cups or pockets of the palms. A hand that does not look capable of clasping is badly drawn. Study your own hands.

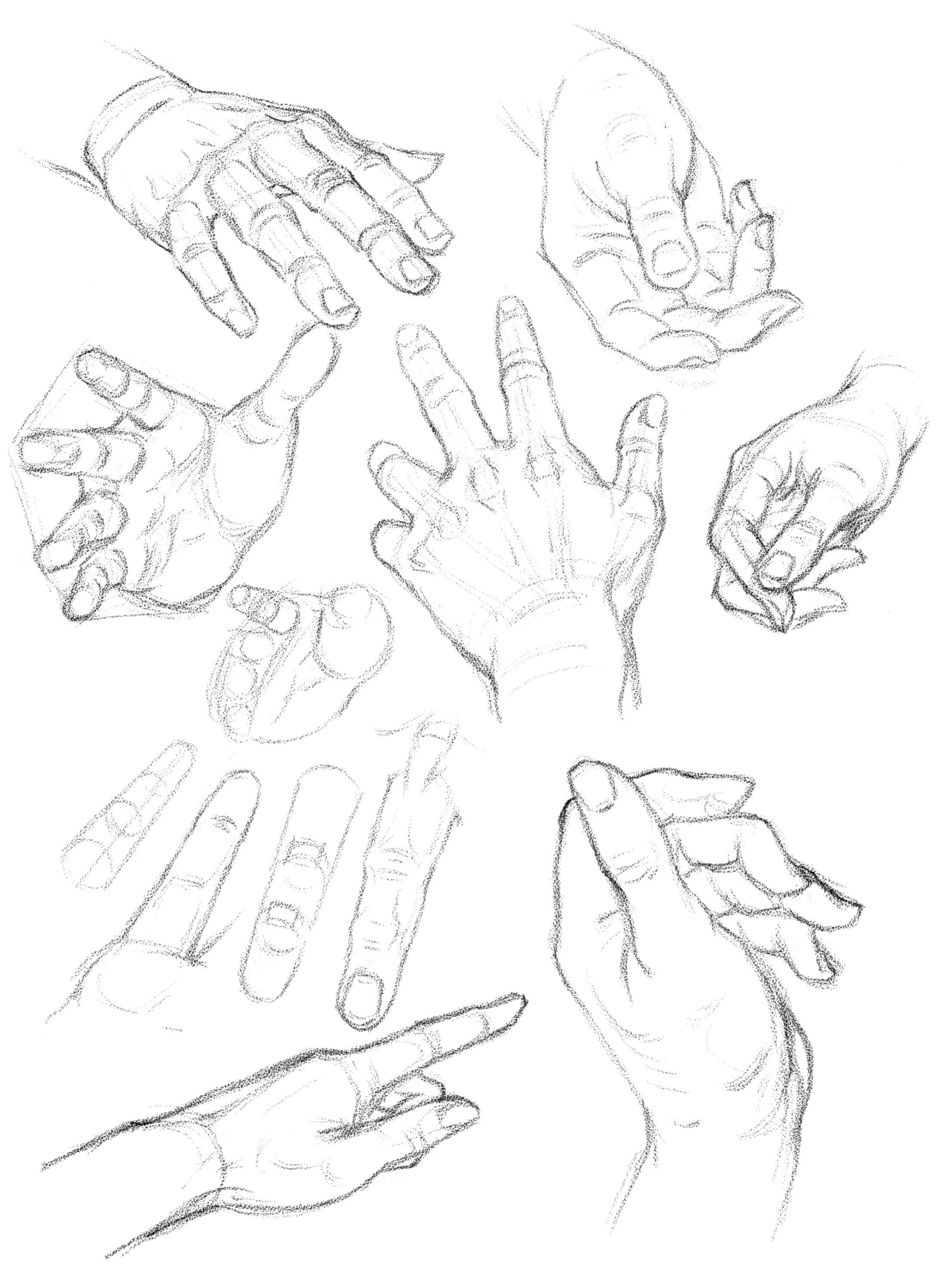

PLATE 82. Foreshortening in drawing hands

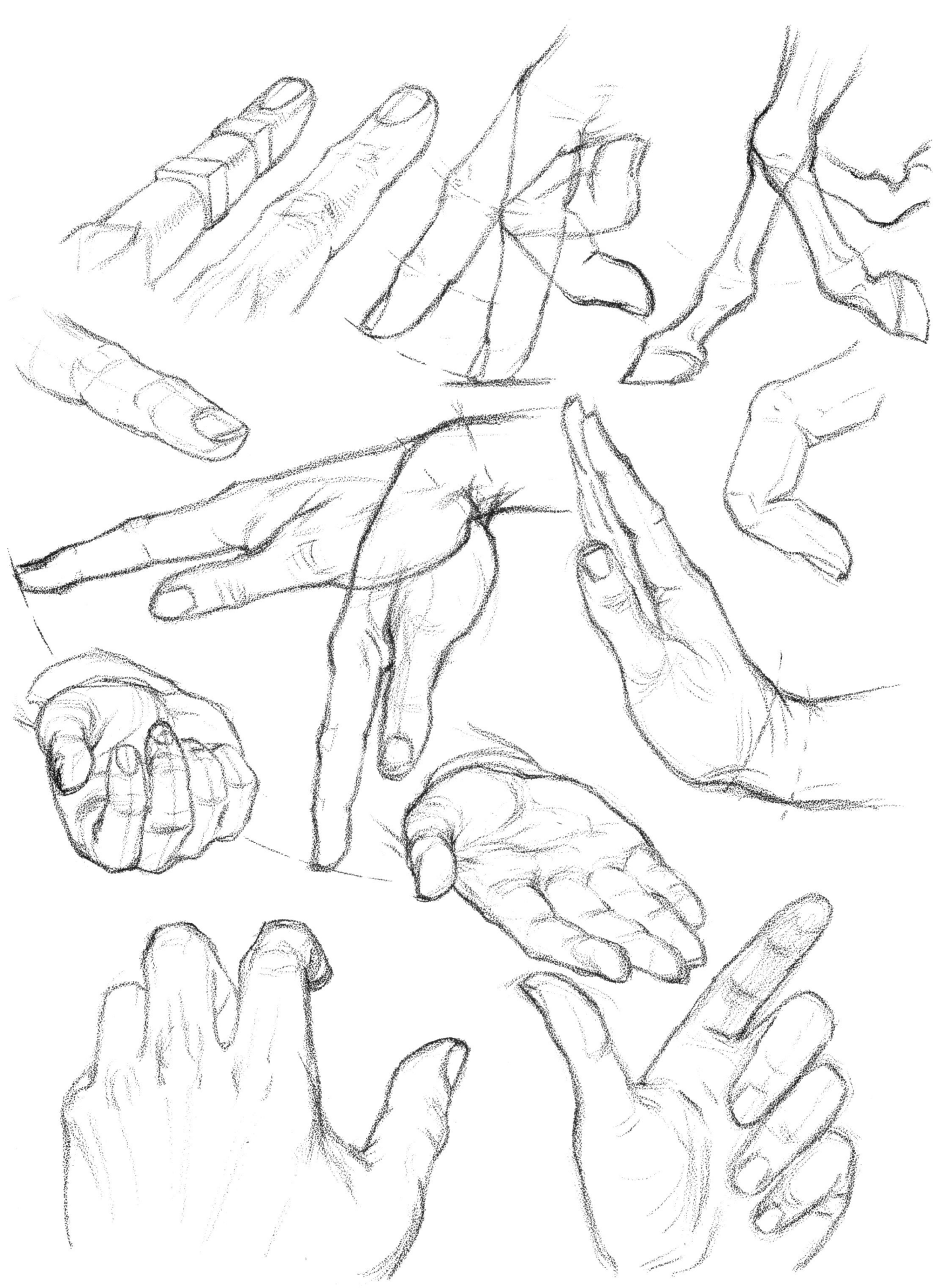

PLATE 83. The hand in action

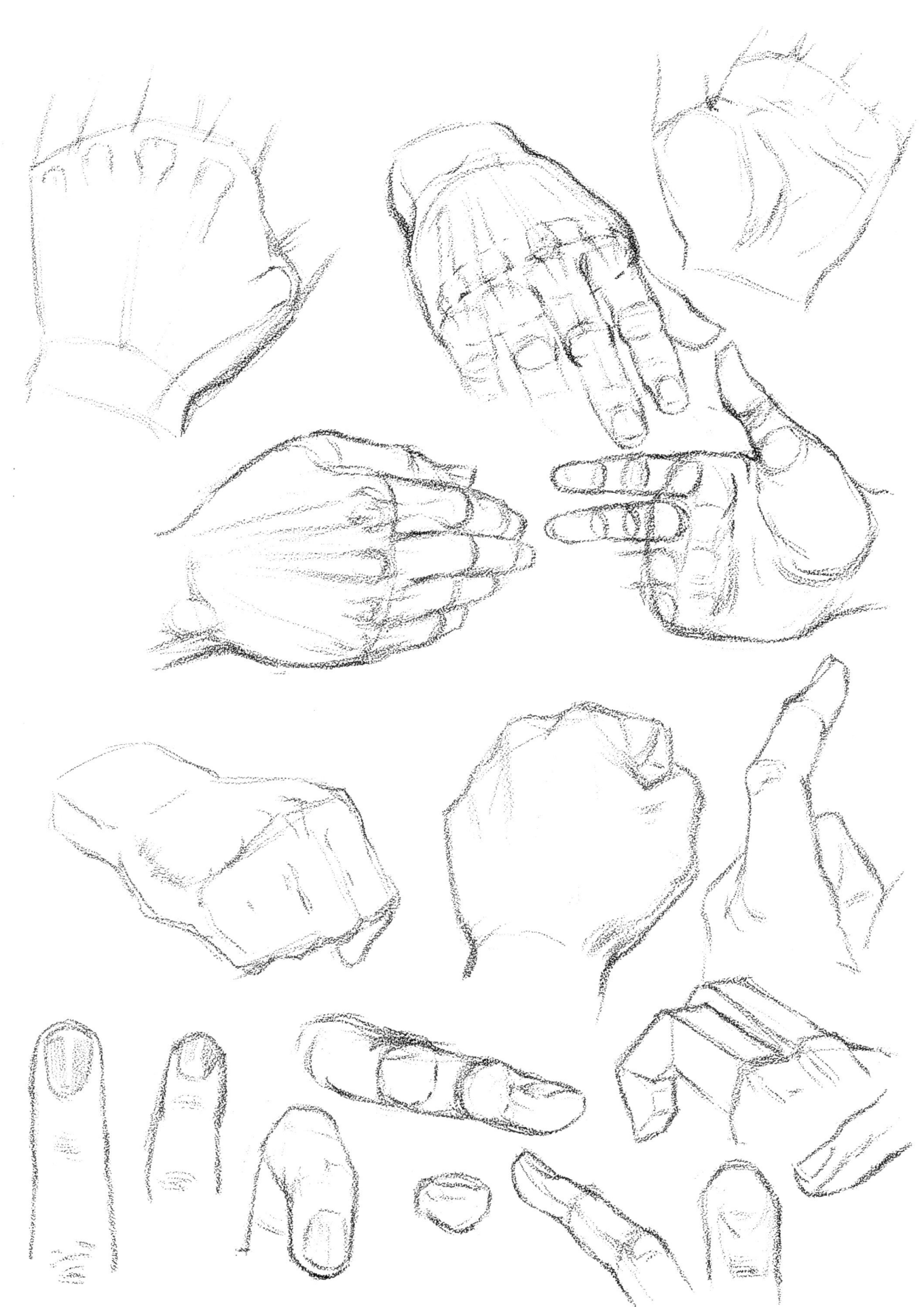

PLATE 84. Knuckles

142

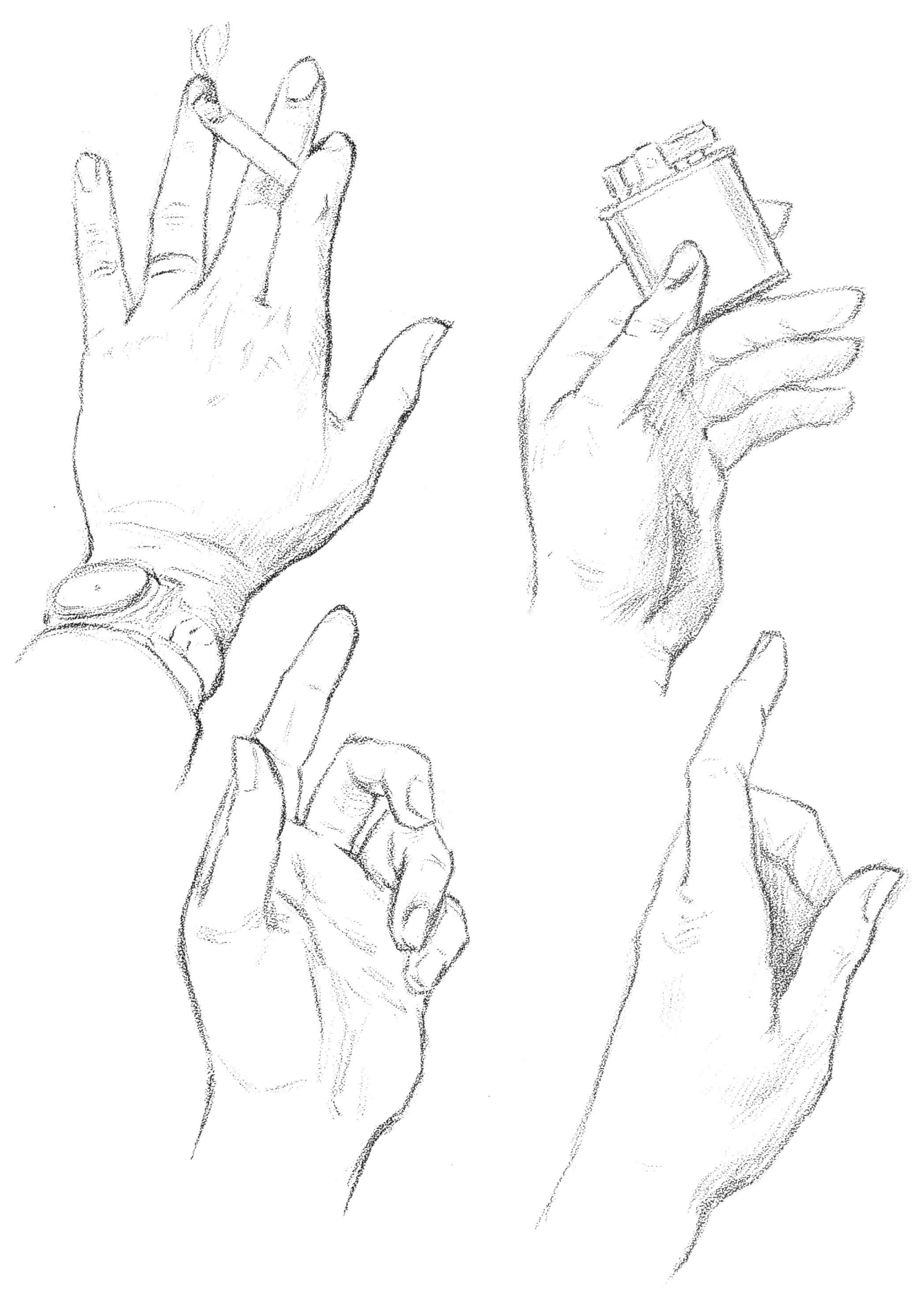

PLATE 85. Drawing your own hand

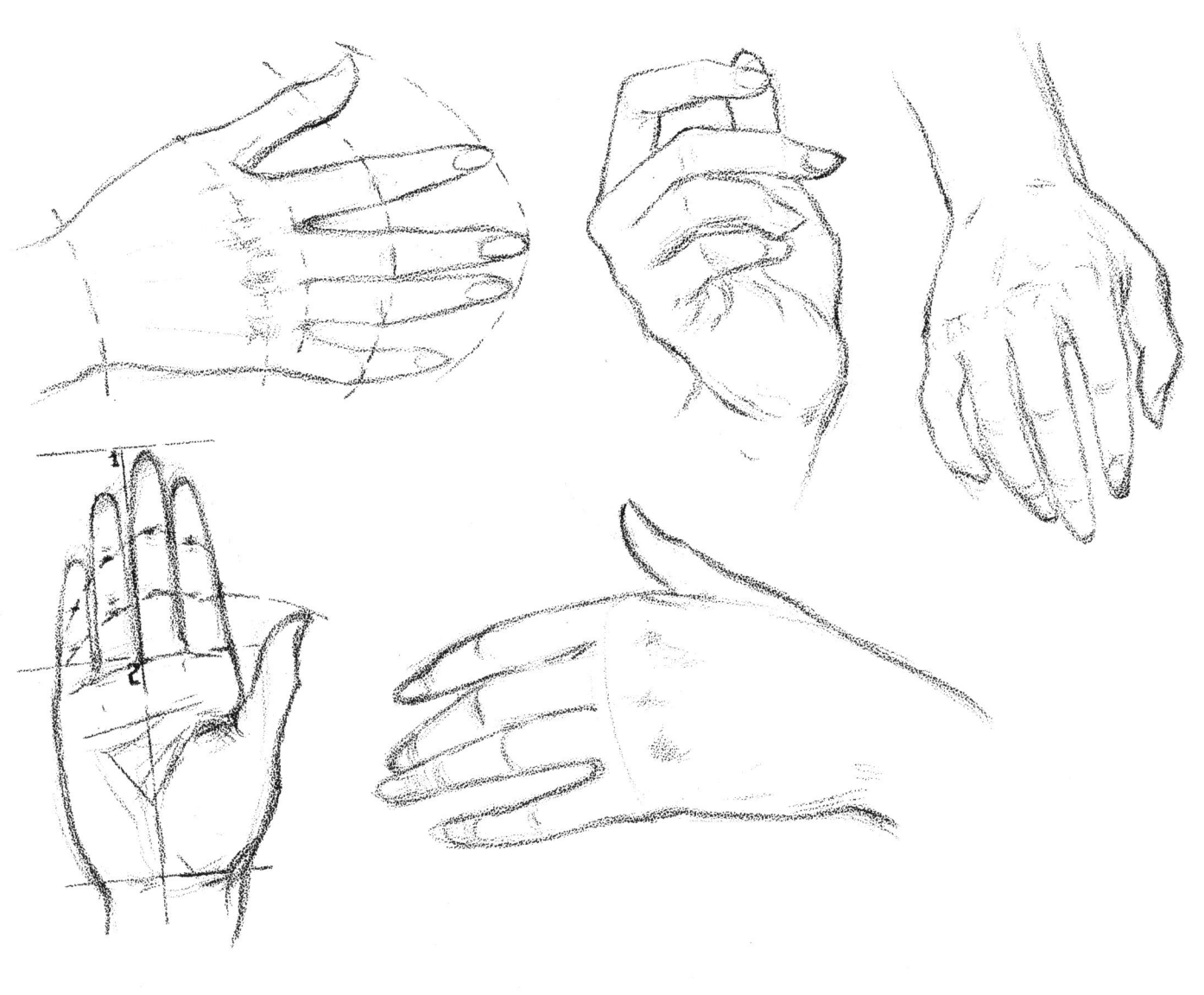

PLATE 86. The female hand

Women's hands, like their faces, differ from those of men chiefly in having smaller bones, more delicate muscles, and generally more roundness of planes. If the middle finger is made at least half the length of the hand on the palm side it will be more graceful and will characterize the hand as feminine. Even though feminine hands are slim, they still have amazing tenacity of grip. The long fingernails, oval in shape, add charm.

144

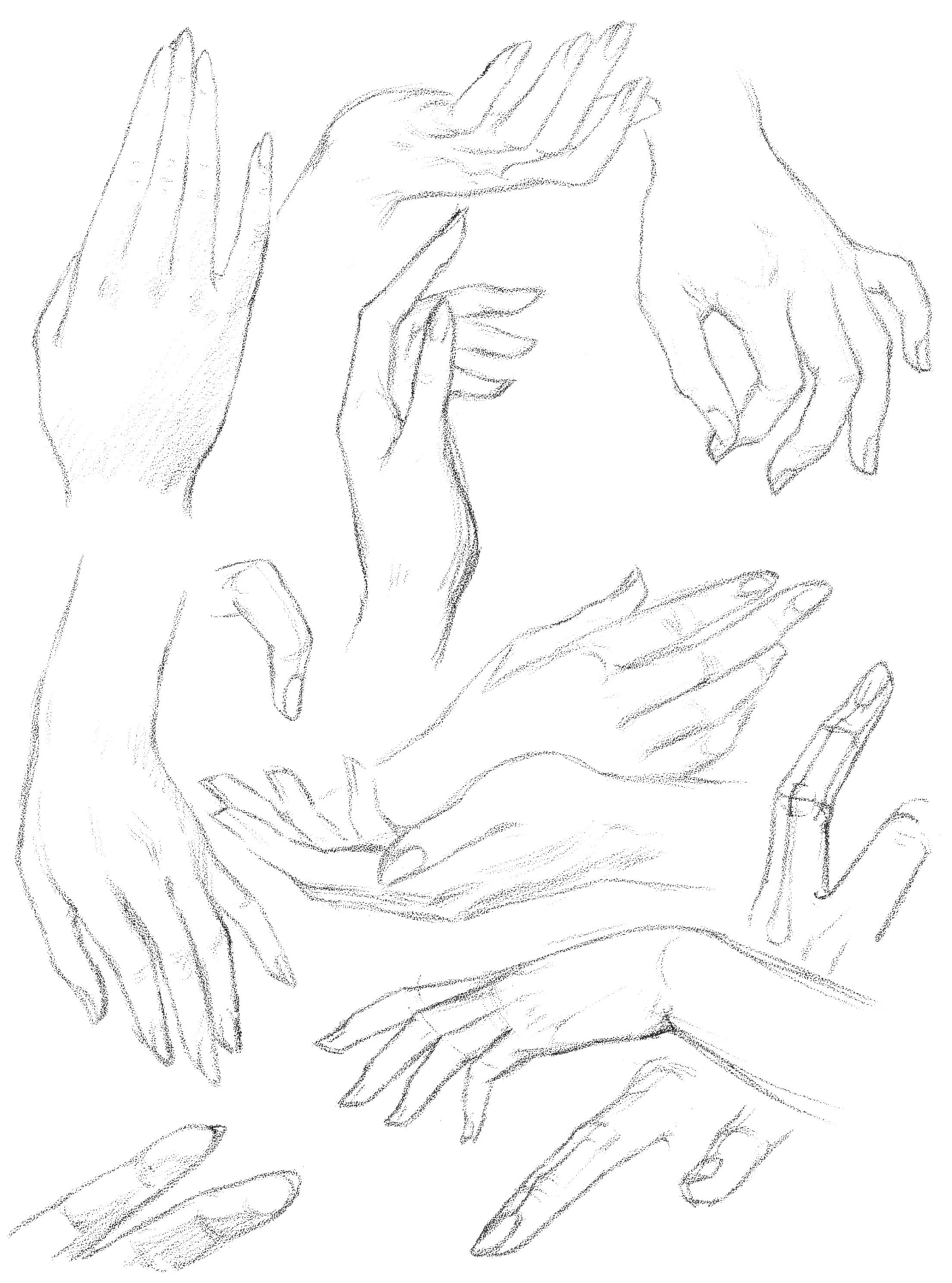

PLATE 87. Tapered fingers

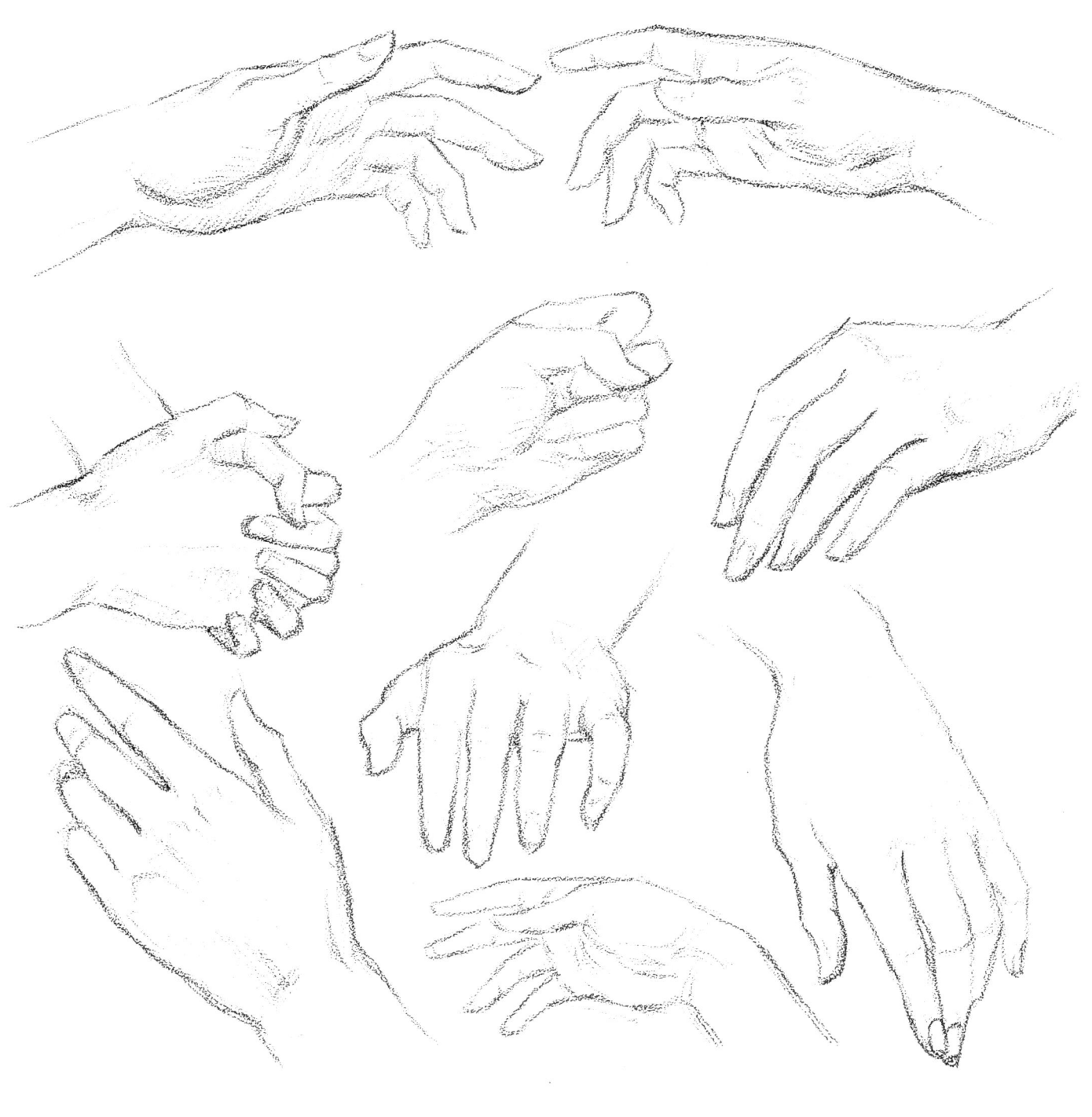

PLATE 88. Make many studies of hands

There is only one sure way to learn to draw hands, and that is to draw many, many studies. With hands, more than with anything else, proper spacing is essential. You must fit the fingers onto the palm in the particular view you see before you. Hands are almost never straight and flat. Judge the spaces between the knuckles carefully. Much of the time the view will require foreshortening, as shown in Plates 82 through 85.

146

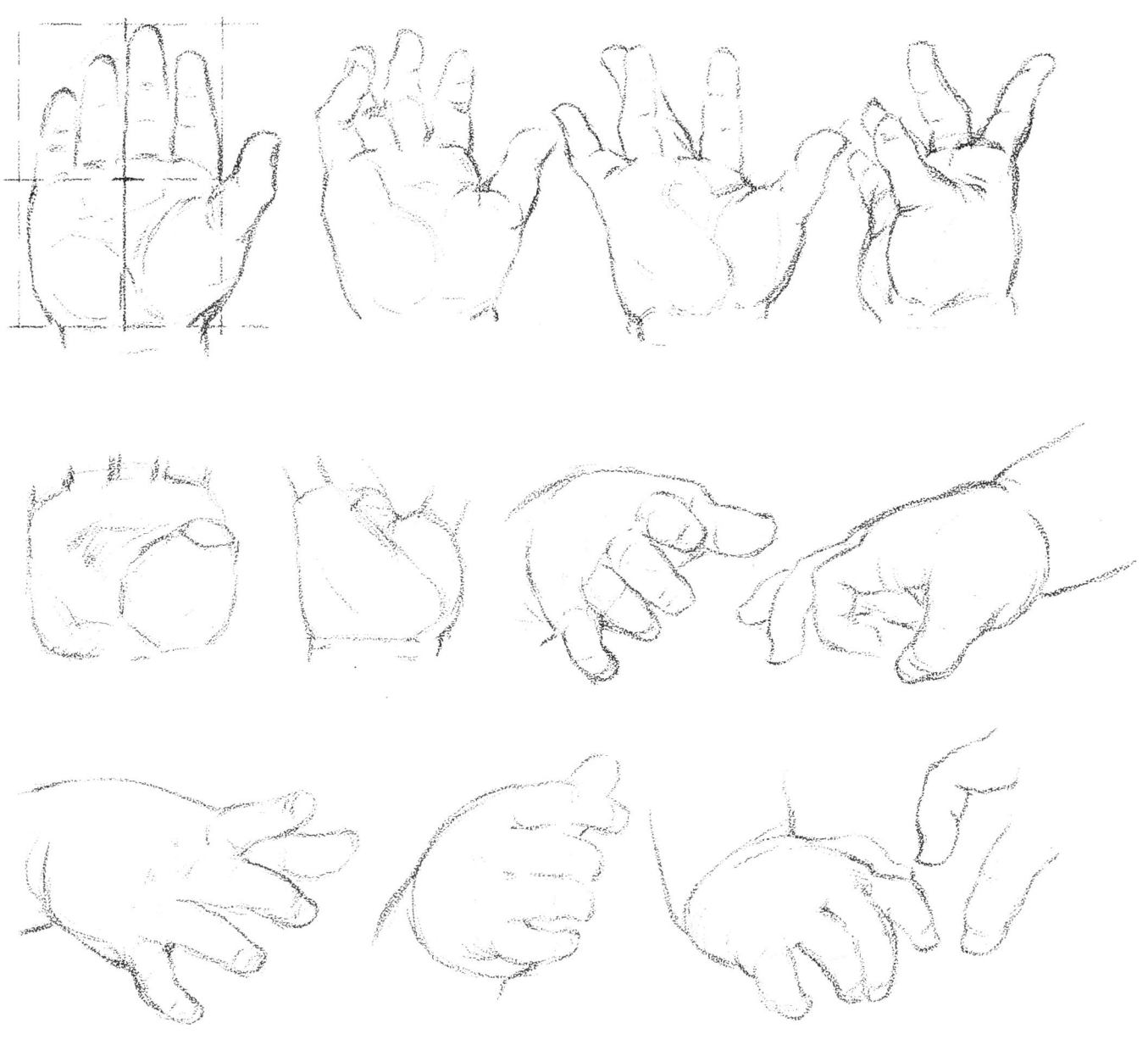

PLATE 89. The baby hand

Babies' hands are a study in themselves. The basic difference from adults' hands is that the palm is relatively thicker in relation to the small fingers. The thumb muscle and heel of the baby hand are proportionately very powerful. Quite young babies have a grasp equal to their own weight. The knuckles across the back of the hand are buried in flesh and are indicated by dimples. The base of the hand may be entirely surrounded with creases. The heel of the hand is much thicker than the pads across the top of the palm.

147

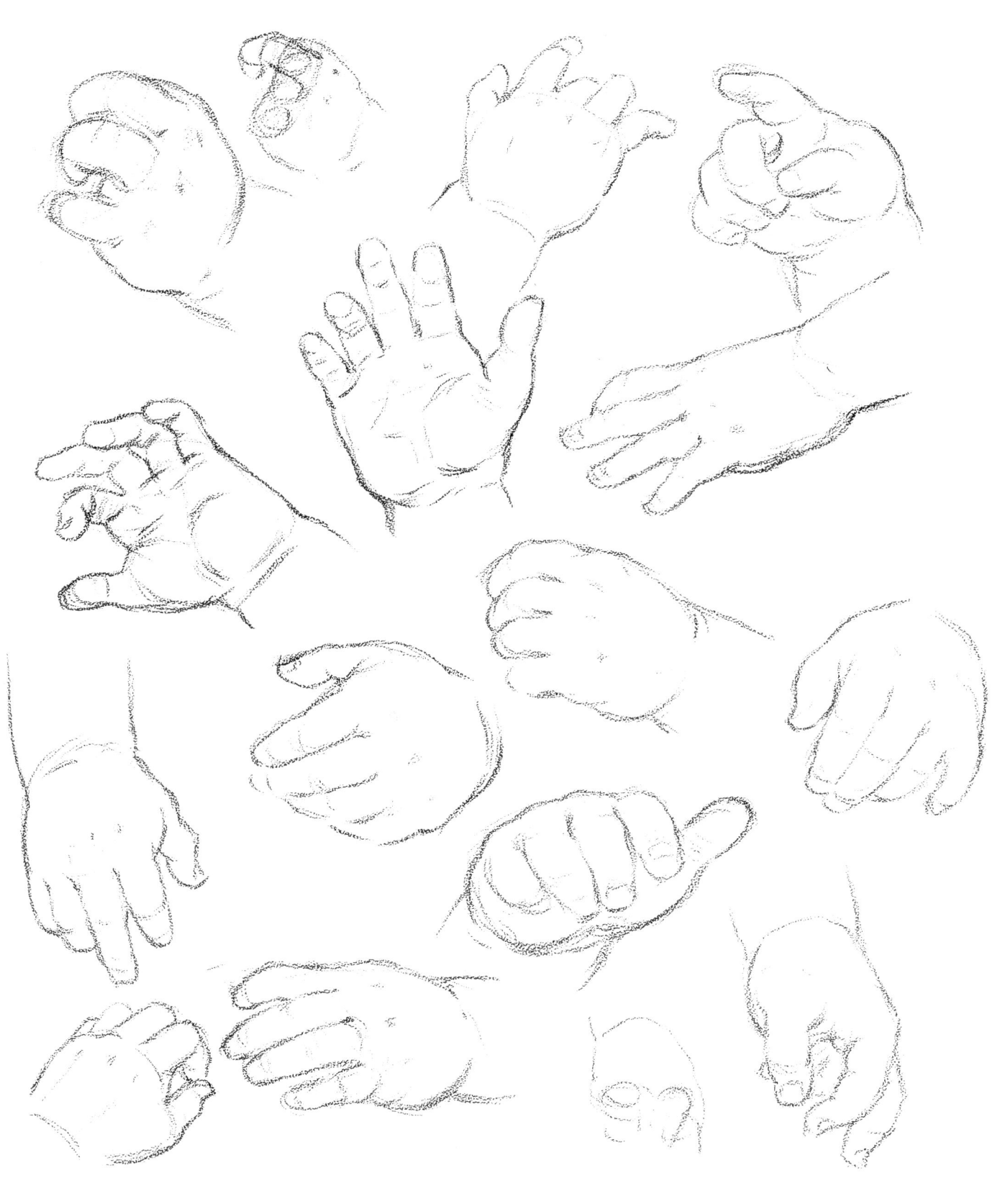

PLATE 90. Studies of baby hands

148

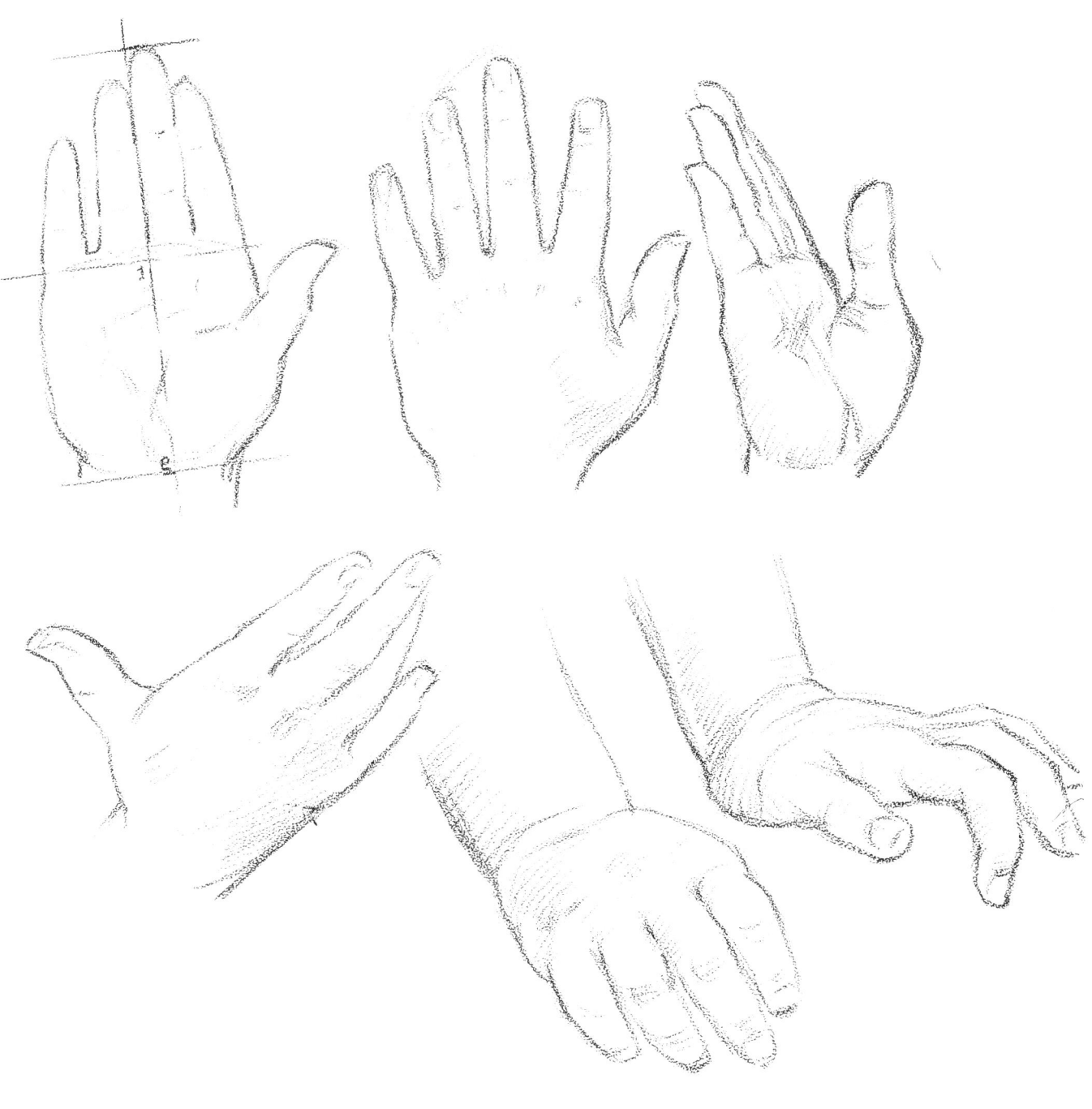

PLATE 91. Children's hands

The child's hand is halfway between that of the baby and that of the
teen-ager. This means that the thumb muscle and the heel of the hand
are thicker proportionately than they are in the adult hand, but not as
thick in relation to the fingers as they are in the baby hand. The fingers in
relation to the palm are about the same as in the adult. The whole hand
is smaller, a little fatter, and more dimpled, and the knuckles are of
course smoother.

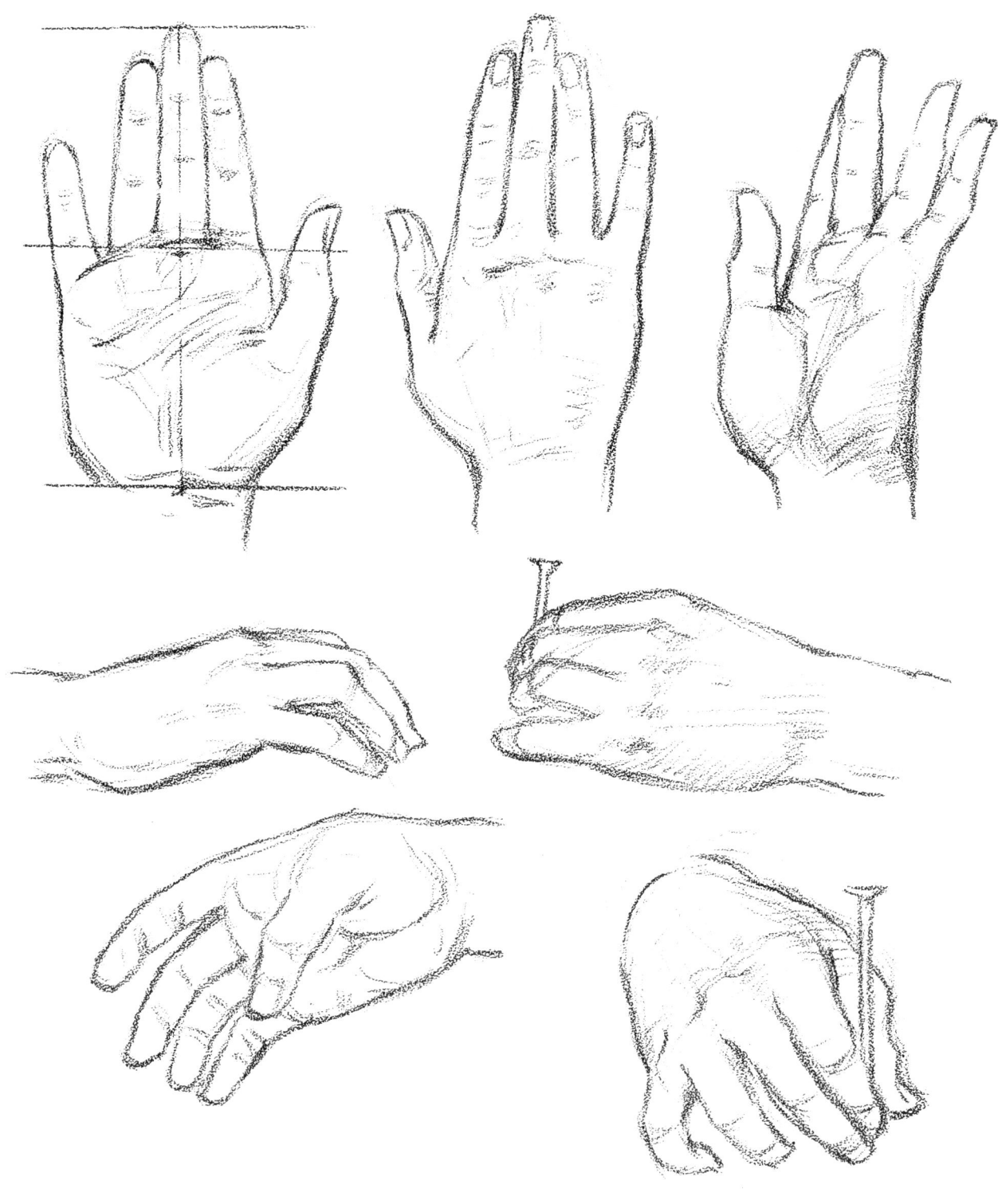

PLATE 92. The proportions remain fairly constant

At grammar-school age there is very little difference between the hand of a boy and that of a girl but at adolescence there is a big change. The boy's hand is much larger and sturdier, showing development of bone and muscle. The girl's hand never develops the big knuckles of the boy's, since the bones stay smaller. The heel of the hand develops in the boy, but stays much softer and slimmer in the girl. In the boy's hand the fingernails as well as the fingers are slightly broader.

150

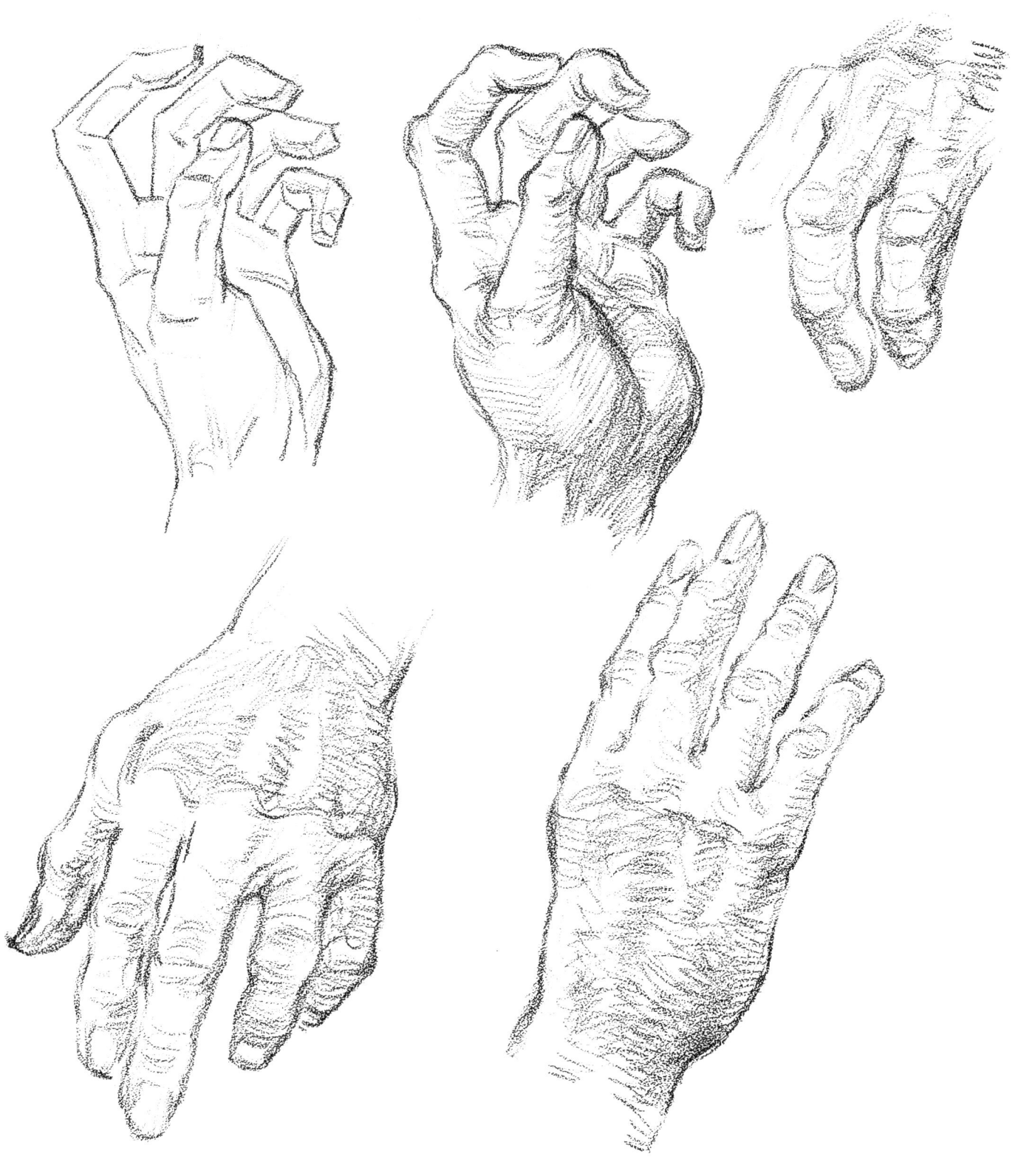

PLATE 93. The hand ages

Once you have mastered the construction of hands, old people's hands are a delight to draw. Actually they are easier than young people's, since the anatomy and construction are more obvious and show clearly on the surface. While the basic construction is the same, the fingers get thicker, the joints larger, and the knuckles protrude. The skin becomes wrinkled, but this need not be emphasized except in a close-up view.

151

A Farewell to the Reader

IN CONCLUDING this book, I want to thank the readers of my previous books for their very kind letters. Because of the large number of these, and because of the pressure on my own time, I have never been able to answer as many as I wished to. If my books have helped you, I am happy.

It is only within the past decade that so many books on drawing and painting have been available. Perhaps another seems superfluous, but in investigating before starting this one, I found very few which concentrated on heads or hands. Both are so important to commercial and portrait artists that I have undertaken to fill the gap. It is my conviction that such a book should come from a person whose livelihood has depended upon the very material he is writing about. In this capacity I have felt that I could substitute actual practice for theory, because my own work based on the principles given here has proved itself by actual sales to leading publications over a long period of time.

There are many fine men in the field of commercial art, and many fine teachers in the schools, who would be capable of handling the same subject. It is largely a matter of finding the time and energy for such an effort in an already full schedule. I have found, however, that time can be apportioned for almost any endeavor that is interesting and pleasant to undertake, simply by curtailing competing pleasures. Much of this book has been done in the evenings or at times between the pressure of other work. My hope is that if I could find time to do the book, others could also in the same way set aside time to study it. My end of the effort is completed, but I am still concerned that it will go out and do the job for young people that I want it to do.

The men in the field who are now the greatest contributors are men who had to come up the hard way, without much knowledge available in books, grasping here and there for information together with much personal practice and experiment. Books will not do the work for anyone, but they can make individual effort more practical and profitable, speeding the acquiring of much-needed knowledge, so that the artist can have more years of successful practice.

It is not my intention to have my readers stop their study of the head and hands with the closing of this book. My aim has been to help them to a well-grounded start that will give their own ability the best of chances. We know that a head cannot be well drawn by any approach that does not, in the final effort, produce solidity and good construction. The portrayal of character must come from specific analysis and from understanding the general anatomy of the head. If I have shown you how that analysis can be made and the reasons for the things that happen in drawing a head, your own progress will be greatly accelerated.

Aside from technical knowledge, I feel that the artist must have a certain reverence for the beauty of the construction of the head, the qualities of its forms that give it individuality, plus a desire for beauty of craftsmanship in the rendering. He should strive never to let his technique become a routine formula, by which all heads are done in the same manner. Let him experiment constantly with the expression of his basic knowledge. Some heads can be done best by suggestion, others by complete detail and fidelity to life. Some will be more interesting if rendered in line, others by tonal suggestion. The result should never look as if it came off an assembly line. To vary your technical style is not easy; neither is keeping your thinking varied. A great deal of practice and experiment is required.

A very fine idea is for a group of young artists

to organize a sketch class, meeting once a week, sharing the cost of a model and other expenses. Such a class offers each man the possibility of learning from the others, and it also establishes friendships which last a lifetime. We did this in my early days in Chicago. Many of the men in that group have forged ahead in their fields, and some are doing the outstanding work of the country. While each must be credited with a great deal of individual effort, there is no doubt that all gained from the collective experience. Of course, any person intending to make a living at art should attend a good art school if possible. But training need not stop there. In the group

I mention, all the fellows had finished their academic work and already were active in the field, but they were all interested in learning more and so organized this informal clinic.

I have enjoyed the preparation of this volume, even if it turned into a mountain of work. I wish every reader the best of luck, and I hope that each will find something in these pages that will be of lasting value. For those to whom drawing is a hobby rather than a profession, I hope the simplification of their problems will bring them still greater happiness in their chosen pastime.

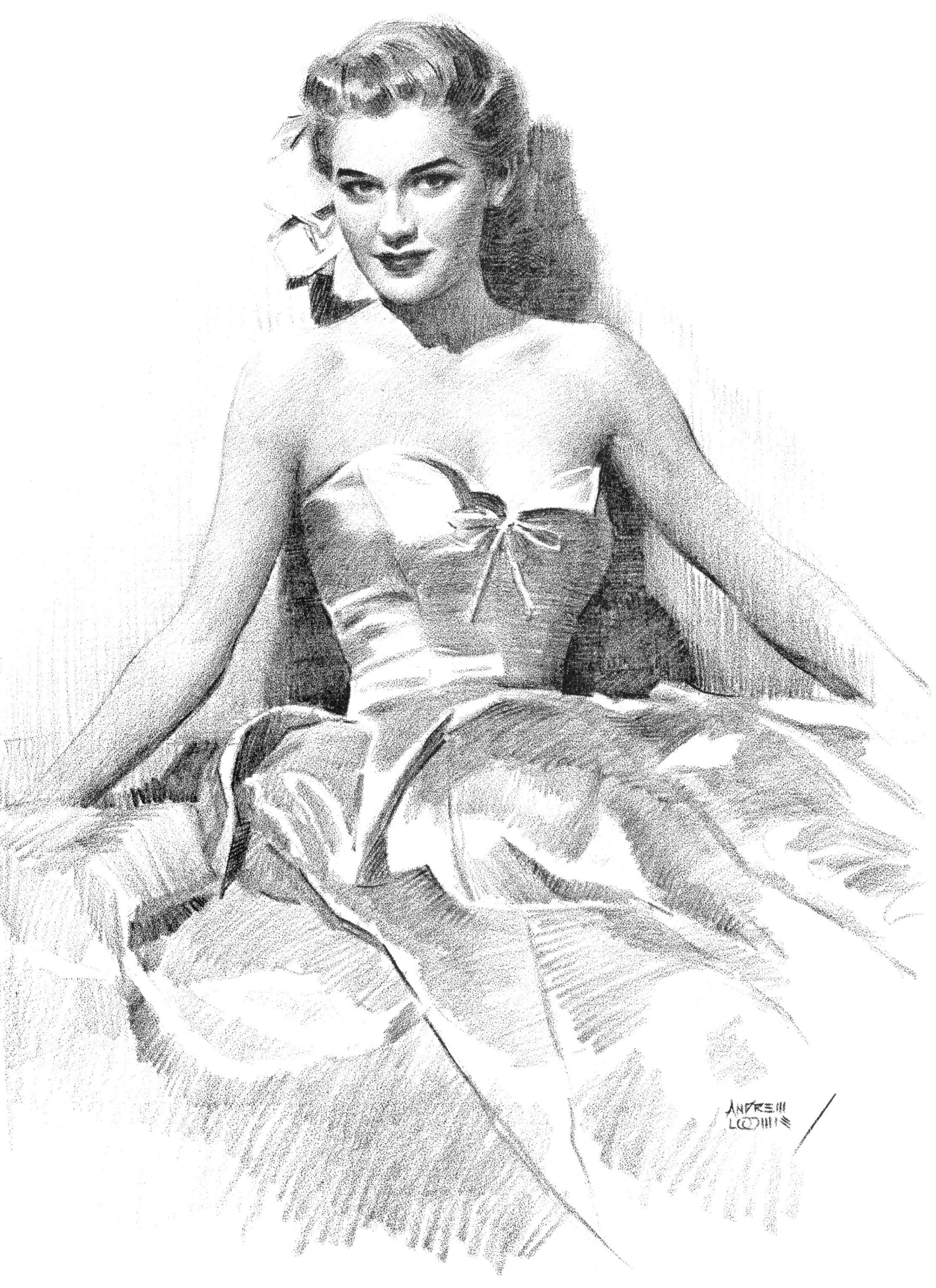